GRAND ILLUSION

Other Books by Wayne Barrett

Rudy! An Investigative Biography of Rudolph Giuliani

Trump: The Deals and the Downfall

City for Sale: Ed Koch and the Betrayal of New York

Other Books by Dan Collins

I, Koch

The Millennium Book

In the Name of the Law

GRAND ILLUSION

THE UNTOLD STORY OF RUDY GIULIANI AND 9/11

WAYNE BARRETT AND
DAN COLLINS

Research Assistance by
ANNA LENZER

HarperCollinsPublishers

 For information, address HarperCollins Publishers, 10 East 53rd Street, New York, NY 10022.

HarperCollins books may be purchased for educational, business, or sales promotional use. For information, please write: Special Markets Department, HarperCollins Publishers, 10 East 53rd Street, New York, NY 10022.

Library of Congress Cataloging-in-Publication Data is available upon request.

ISBN-10: 0-06-053660-8
ISBN-13: 978-0-06-053660-2

06 07 08 09 10 NMSG/RRD 10 9 8 7 6 5 4

For Jack Newfield, who loved his city and his craft.

WAYNE BARRETT

For the men and women who sacrificed so much in the rescue and recovery effort at Ground Zero.

DAN COLLINS

CONTENTS

PART ONE SEPTEMBER 11

PART TWO THE UNLEARNED LESSONS OF 1993

PART THREE THE AFTERMATH

PART FOUR 9/11 REVISITED

GRAND ILLUSION

PART ONE

SEPTEMBER 11

CHAPTER 1
THE WORLD TRADE CENTER ATTACKED

FIVES, A RESTAURANT in Manhattan's Peninsula Hotel, was one of Rudy Giuliani's regular places, and when New York City's mayor arrived there for breakfast on September 11, 2001, his favorite table was waiting for him. It was large, round, and located in a nook beneath a bay window. As always, the tables in front of and behind him were left open. The seat Giuliani selected gave him a view of Fifth Avenue, the entryway, and a good portion of the restaurant itself. Although he didn't like to be disturbed while dining, Giuliani always seemed to have an eye on what was going on around him.

His breakfast companions were Denny Young, a top aide, and a friend, Bill Simon, who was hoping to run for governor of California with Giuliani's endorsement. Entering the restaurant, Giuliani worked the room, smiling and shaking hands before taking a seat at his table. The mayor's security detail split up as he sat down. All modern New York mayors have traveled with a retinue, and Giuliani's concern for physical protection was long-standing, the product of an earlier career spent prosecuting Mafia cases. One bodyguard took up position at the hostess station at the head of the stairway leading into Fives. The other stood in front of the wall behind the mayor's table.

Zack Zahran, the restaurant manager, watched his celebrity guest as the three men ordered coffee and began discussing Simon's gubernatorial campaign. At around 8:50 A.M., he saw one of Giuliani's bodyguards leave her post near the mayor's table and come forward to whisper in the mayor's ear. Zahran saw no change in Giuliani's expression or sense of emergency in his demeanor. As he recalled it, the mayor chatted with Young and Simon for another minute or so before exiting the same way he arrived—moving through the restaurant for another round of smiles and handshakes.

Denny Young followed his boss. Left behind was Bill Simon. According to a Simon aide, Giuliani told his friend, "A plane hit the World Trade Center. I've got to go," and Simon replied, "All right." In Giuliani's subsequent account of his departure, Simon came out looking more prescient. "Without knowing the enormity of what had happened," the mayor recalled, "Bill said to me, 'God bless you.' "[1]

Also left behind on the table were three unopened menus. As the day unfolded and images of the crashing towers and a soot- and ash-covered Giuliani flashed on TV screens, manager Zahran had the same thought over and over again: "Oh my God, the man didn't have breakfast!"[2]

It was the beginning of the most important day in many American lives, Rudy Giuliani's included. Later, when the chorus of praise for Giuliani's performance would swell so loud the mayor of New York City began to sound like a combination of Winston Churchill and Spiderman, his political peers began to grumble that he had only done what any responsible elected official would have done in his shoes. Mark Green, the leading Democratic candidate to replace Giuliani in the 2001 election, said as much at the time. "I actually believe that if, God forbid, I had been the mayor during such a calamity, I would have done as well or better than Rudy Giuliani," he said, and was hit with a wave of outrage from New Yorkers who wanted to believe that Giuliani was every bit the unique hero he had seemed that day.

We will never know how Green would have behaved as mayor under any circumstances—he lost the election to Michael Bloomberg that November. But on September 11, no other public figure rose to the occasion the way Giuliani did. It took George W. Bush more than a day to completely digest what was going on and to craft an appropriate response. The president was, of course, operating in a different environment. Bush had trouble getting a full picture of what was happening—the high-tech Air Force One kept losing telephone and television reception.

Giuliani, on the other hand, began to understand that things were very, very bad a few minutes after he left the restaurant. He, Young, and two police bodyguards sped downtown in a Chevrolet Suburban, and as the SUV passed through Greenwich Village, the mayor observed doctors and nurses in operating gowns standing on the street, outside St. Vincent's Hospital. He knew then, he said later, that it "had to be even worse than I thought." And it was getting far more disastrous by the moment. A little more than 16 minutes after the first jet hit the North Tower, a second plane, United Airlines Flight 175, struck the 78th through 84th floors of the South Tower.

Giuliani, whose car was about a mile away from the World Trade Center when Flight 175 hit, saw the explosion but assumed it was coming from the wreck in the first building. "And then I was informed within about 30 seconds that a second plane had hit the World Trade Center," he said. "At that point, we knew there was a terrorist attack going on."

Inside the North Tower of the World Trade Center, above the floors where a jet plane filled with fuel had just crashed, brokers and secretaries and other workers were calling their families worriedly, still sitting at their desks and totally unable to comprehend what was happening to them. Mike Pelletier, a commodities broker who worked on the 105th floor, called his wife, Sophie, in Connecticut. "He just said, 'Soph, an airplane just went through the building. I don't know what we're going to do.' He said he loved me," she recalled later. "And it took me a second to just realize what was happening. I said, 'Oh my God, is there help?' He said, 'We don't know. We don't know. We can't tell.' " Mrs. Pelletier called 9-1-1 and got emergency response in Connecticut, where the operator laughed, unbelieving. There would be no help for those above the impact of either plane, except for 18 people in the South Tower who found a passageway down.

Workers on the buildings' lower floors were taking control of their own fate and heading for the stairway. Eric Levine, a Morgan Stanley employee whose office was on the 64th floor of the South Tower, fled immediately after the first plane struck the North Tower. He had reached the 50th or 51st floor when his own building was hit. A tremendous explosion knocked him down a flight of steps. "I then tried to stand up but the building was still shaking and the lights were flickering on and off. It was terrifying! Then the building began to sink. That's the only way I can describe it. The floor began to lower under my feet and all I could think about was that it would crack open and I would fall hundreds of feet to my death," he recalled.[3] Out of the darkness came screams, shouts, and prayers. Finally, the building settled and the evacuation resumed with the panicked flight of people down the stairway. Levine waited against the stairwell wall for the crowd to calm down and then resumed his own descent. Just before he made his escape from the building, he looked through a window into the plaza between the two towers. There were bodies scattered everywhere, some still smoldering.

No one knew it at the time, but of the 17,400 occupants of the building that morning, roughly 15,000 would survive. Only 118 of the approximately 2,150 who died were occupants of floors below the impact of the planes.[4] Survival was mostly a matter of place and time, and was determined more by what floor you were on when your 110-story building was struck than by any other factor.

THE EMERGENCY PERSONNEL racing to the World Trade Center from all around the city had no way of knowing that their heroism would, in many cases, end not in saving civilian lives but simply in placing themselves in mortal danger. They knew only that their job was to run toward the things normal people fled—fires, shootings, collapsing buildings. And so they came, racing to the trembling towers and the falling debris. And with them came Rudy Giuliani.

The mayor's original destination was the much-ballyhooed command center he had built in the shadow of the Twin Towers. But the elaborate bunker—constructed to deal with just such an emergency—was almost empty when he arrived. Giuliani then began a harrowing trek to find a temporary headquarters where the city could manage the unfolding disaster. It was a march that would help to transform him into a national hero. Dodging debris, walking calmly uptown through air so filled with dust and ash that people could not see the pavement at their feet, he was the father figure the city needed on a day when every New Yorker felt a little lost and frightened.

George Bush received word at around 8:55 a.m. that a plane had hit the World Trade Center. He had already arrived at an elementary school in Sarasota, Florida, to watch a group of second-graders read, and he decided to go ahead with the photo op. "I thought it was an accident," Bush later recalled. "I thought it was a pilot error. I thought that some foolish soul had gotten lost and—and made a terrible mistake." He continued with the planned event, and was listening to the children read a story about a pet goat when his chief of staff, Andrew Card, stepped into the classroom around 9:05 a.m. "A second plane hit the second tower. America is under attack," Card whispered in Bush's ear. A look of panic crossed the president's face. Later, he remembered thinking, "I have nobody to talk to. My God, I'm Commander-in-Chief and the country has just come under attack!"[5]

Nevertheless, Bush remained in the classroom for another seven or eight minutes after learning that a second plane had plowed into the South Tower. As a sympathetic writer later described the scene in Sarasota: ". . . Without all the facts at hand, George Bush had no intention of upsetting the schoolchildren who had come to read for him. The rest of the children's story about the goat did not register with him at all, but the president, raising his eyebrows and nodding, interrupted the second graders to praise them. 'Really good readers, whew!' Bush told the class. 'This must be sixth grade.' "[6]

When Rudy Giuliani got word of the crash, shortly before Bush, he left the restaurant far faster than the president left the classroom. On the northern edge of the 16-acre World Trade Center, Giuliani met his police commissioner, Bernard Kerik, and other top city officials at about 9:07. They decided to walk south from Barclay Street to a command post set up by the Fire Department near the burning buildings. Looking up, Giuliani saw a man lean out of a window "about the 102nd floor of the Tower," and leap into the air. "I saw him jump and followed his whole trajectory as he plummeted onto the roof of 6 World Trade Center," Giuliani recalled. "I looked up again and saw other people jumping. Some appeared to be holding hands as they plummeted. They were not blown out of the building. They made a conscious decision that it was better to die that way than to face the 2,000-degree heat of the blazing jet fuel."

NO OTHER ASPECT of the unfolding tragedy was more disturbing than the sight—and sound—of people jumping to their death from the Twin Towers. To Stephen King, one of the fire chiefs supervising the evacuation of the North Tower, the bodies crashing into the roof over the lobby of the tower came with the rhythm of bursting popcorn kernels. The thudding noises were utterly unnerving to the chiefs in the lobby and made it difficult for them to think clearly as they formulated a plan of action. "It was unlike anything I had ever witnessed in my life, or even thought was possible," King recalled. "Every time I heard a body hit that roof, it sent chills through my body."

September 11 was King's first day back on the job after a long leave he had taken to be with his wife, who had contracted a rare and deadly form of breast cancer. His office at the Brooklyn Navy Yard had a great view of the World Trade Center. King didn't actually see Flight 11 plow into the North Tower at 8:46 A.M., but within a minute of the collision, he and his driver were headed for the WTC. Like hundreds of other firefighters around the city, King didn't wait for a "ticket," or formal order, to race to the burning building. Within eight minutes, his car pulled up next to the burning tower. Too close, as it turned out. Falling bodies and debris rained down. The bodies, traveling at well over 100 miles per hour, exploded all around them. King's driver jammed the car into reverse. The two men took shelter under a scaffold and then looked up at the tower in order to time their dash into the lobby.

* * *

RUDY GIULIANI'S SMALL party was walking through the falling ash like characters in some ancient epic. They reached the Fire Department's command post at West and Vesey Streets, which was under a shower of debris falling from the flaming towers. There, the mayor was briefed by the brass. Chief of Department Pete Ganci told him that firemen were ascending the staircases in both towers to assist workers fleeing the buildings.

"We can save everybody below the fire. Our guys are in the building, about halfway up the first tower," Ganci said. The mayor realized that Ganci was also sending a second sobering message: everybody above the fire was doomed. The important thing, Ganci told the mayor, was for all of the survivors to head north, away from the towers, as quickly as possible. That message became Giuliani's mantra, which he would repeat again and again. The mayor decided to push north himself, walking back toward Barclay Street to try to set up a command location in an office building selected by Kerik. Before leaving, he bid goodbye to Ganci, Chief of Special Operations Ray Downey, First Deputy Commissioner William Feehan, and the Fire Department chaplain, Father Mychal Judge. All four would die that day.

"IT WAS PRETTY clear to us that there was no way to put out a fire of this magnitude," Stephen King recalled. "Our concern was the need to evacuate the building in an orderly manner." The Fire Department had known for years that extinguishing a major fire in a high-rise building was a practical impossibility. "The best kept secret in America's fire service," wrote Vincent Dunn, the deputy chief of the New York City Fire Department in a 1995 trade magazine article, "is that firefighters cannot extinguish a fire in a 20 or 30,000 square foot open floor area of a high-rise building." The World Trade Center had floors of 40,000 square feet—almost a clear acre apiece.

None of the chiefs gathered in the lobby had discussed a building collapse, but the possibility played in the back of King's mind as he thought about the heat of the fire weakening the steel beams that supported the tower. Nevertheless, he joined the upward surge of firefighters, hoping to evaluate the progress of the evacuation. King walked up about eight floors and was well pleased with what he found. Office workers were making their way down three sets of stairs in an orderly manner. There was no panic, nor did he see any signs of smoke or fire. Some of the workers rushing down the stairway reached out to touch the shoulders of the firemen racing up the stairs. "They couldn't believe that the firemen were actually going up. It was a memory that will stay with me always," King recalled.

King tried to radio his report to the chiefs in the lobby below, but reception in the high-rise building was poor. He realized he would have to return to the lobby to brief the other chiefs on the evacuation. In his almost unique case, the defects in the Fire Department radio system proved to be a lifesaver. Had King been able to reach the chiefs by radio, he would have continued his climb up the stairs.

GIULIANI HAD BEEN at the scene of the disaster for about 40 minutes when, at around 9:50 A.M., he commandeered a small office building at 75 Barclay Street, where he hoped to establish a temporary command post. Cell phone communication had become nearly impossible, and Giuliani used the landline phones at the Barclay Street building to contact the White House, which was being evacuated.

President Bush had been shepherded into Air Force One, where, surrounded by confused and security-obsessed aides, he wound up circling in the air for about 40 minutes before heading for Barksdale Air Force Base in northern Louisiana. Bush was out of touch with the country, which was waking up to the enormous disaster that was taking place in New York and Washington, and he was unreachable when Mayor Giuliani urgently asked a presidential aide to put him through. The mayor was told that Vice President Dick Cheney would call him back soon on the same phone. It took Giuliani a minute to realize that the phone had gone dead.

At 9:59 A.M., the South Tower collapsed, sending an enormous cloud of smoke, gas, dust, and deadly debris rushing through the streets of Lower Manhattan. The mayor, in his newly established command post at 75 Barclay, heard a loud roar but had no idea what was happening. Chunks of steel and concrete blew out the south-facing windows and buried the building entrance in debris. Rudy Giuliani rushed to the basement.

In the North Tower lobby, Stephen King was reviewing blueprints of the building as the chiefs struggled to get a handle on what systems might still be working on the upper floors. Amid the bedlam and confusion, he overheard a radio transmission: "Oh my God! The tower's coming down!" With no hard information about what was happening, King assumed that the North Tower was coming down on his head. "Oh my God. There is no way I am surviving this one," he thought. In what the 30-year Fire Department veteran believed to be his final moments of life, King worked out an eerily accurate picture of the death suffered by hundreds of his fellow firefighters trapped in the Twin Towers. King had been flabbergasted by the subterranean devastation wrought by the first World Trade Center bombing in 1993, and he now saw himself being driven deep underground through a set of subbasements by forces so violent and powerful that no shred of his body would remain for rescuers to recover.

The roar grew louder and louder. What was left of the massive floor-to-ceiling windows in the lobby exploded inward, sending BB-sized bits of glass flying through the air. Stephen King's world went black.

When he regained consciousness, he was amazed to find himself alive but suffocating. "This is ridiculous," he thought as he gasped and struggled to regain his breath. Despite a badly injured knee, he somehow managed to make it out of the lobby. In a state of shock, King moved from utter blackness into a world of white. Was he walking through a cloud? he wondered. Or was he already dead? In King's ghost world, he could see people running and hear them screaming on the street around him, as bodies and debris continued to fall from the sky. But he felt divorced from the chaos around him: "It was like I was outside looking in."

★ ★ ★

AT BARCLAY STREET, Giuliani and his party began a search for a way out of the building. A series of locked basement doors prevented escape. The mayor's men were growing more frightened and apprehensive, though everybody tried to put on a brave front. Then a maintenance man appeared out of nowhere. He led the party through a basement door into an adjoining building at 100 Church Street. There, things were not much better, and Giuliani soon concluded that they needed to move again to avoid the possibility of a building collapse.

"If I have to die I'd rather die outside than get trapped in a building," he remembered thinking. He needed to find a place "to re-establish city government." Unlike George Bush, who was stuck, incommunicado, in Air Force One, Giuliani was determined to speak to his frightened city.

Reporters are another first responder breed who tend to race toward situations that rational people run away from, and outside the building Giuliani saw some members of the media, including Andrew Kirtzman, a TV reporter and Giuliani biographer. "I grab Kirtzman by the arm and say, 'We're taking you with us.' Some of them look a little stunned. I begin holding an ad-hoc walking press conference in which I tell people to remain calm and go straight north."

If the reporters were stunned, it was because it had been a long time since Giuliani had solicited their presence anywhere. He simply had no use for the people covering him. The relationship wasn't so much bad as nonexistent. But that was just one of the many things September 11 was changing. The reporters now became an integral part of Giuliani's traveling emergency team. The mayor wanted to demonstrate that he was firmly in control despite the catastrophe. His party moved north up Church Street in search of yet another new headquarters. Many of the men who were with him would come to be regarded as heroes in their own right because of their connection to Giuliani and their part in his march uptown—their pictures leaping out of glossy magazines and newspaper profiles, courted for TV interviews and deluged with offers of speaking engagements or consulting contracts. Kerik, the police commissioner, Tom Von Essen, the head of the Fire Department, and Richard Sheirer, who ran the Office of Emergency Management, would, in particular, emerge as the mayor's Three Musketeers—almost as identified with the terrible day as Giuliani was.

Giuliani began a series of "walking press conferences" as they marched uptown. Then, at 10:28 A.M., the North Tower collapsed in a terrifying replay. "Fuck!" yelled a mayoral aide. Everyone, including Giuliani, started running away from the second deadly cloud of ash and debris that had been unleashed.

"Just keep going north," Giuliani shouted.

AIR FORCE ONE did not touch down at Barksdale until 11:45 A.M., bearing the president of the United States. He had been out of sight since making a one-minute statement at the Sarasota elementary school. In the interim, both World Trade Center towers had collapsed and the Pentagon had been attacked. "The American people want to know where their dang president is," Bush complained.

BY THAT TIME, Rudy Giuliani was already a legend in the making. After rejecting several buildings as temporary headquarters because they were close to structures that might themselves become terrorist targets, the mayor and his party broke into a firehouse on Houston Street. There, around 10:57, Giuliani found a phone and spoke to the people of New York City. Pleading for calm, he said, "My heart goes out to all of you. I've never seen anything like this. . . . It's a horrible, horrible situation, and all that I can tell you is that every resource that we have is attempting to rescue as many people as possible. The end result is going to be some horrendous number of lives lost."

Giuliani then went about the business of reestablishing his government. He finally settled in around noon at the police academy on East 20th Street as a headquarters and began to plan for the immediate future. His first concerns turned out to be unfounded. Like almost everyone else in New York, Giuliani expected thousands of injured people to jam city hospitals, but, in fact, the circumstances of the disaster had drawn a fairly clean line between the survivors, who were mainly unharmed, and the victims, who never emerged from the building. In addition to the 2,150 occupants killed in the Twin Towers, the other nearly 600 who died were mainly police, firefighters, and plane passengers, including the hijackers.

AT BARKSDALE, BUSH conferred with Cheney on the phone and issued a two-minute videotaped statement in which he pledged that the U.S. "will hunt down and punish those responsible for these cowardly acts." The president looked nervous and vaguely confused, and the appearance did little to soothe a troubled nation. Bush wanted to return to Washington, but was dissuaded by aides. "We still think it's unstable, Mr. President," said Cheney. At 1:25 P.M., Air Force One took off again, headed for Offutt Air Force Base in Omaha, Nebraska. Once airborne, the president's talk turned tough, according to notes taken by his press secretary, Ari Fleischer. "We're not going to have any slap-on-the-wrist crap this time," Bush said at one point. To the vice president, Bush said, "We're at war, Dick. We're going to find out who did this and kick their ass."

WHILE GEORGE BUSH was making America wonder who was watching the store, Giuliani led a televised news conference at the police academy at 2:50 P.M.. It was a masterful performance that left no one in doubt that New York City, at least, was in strong hands. In response to a question about the number of deaths, he said, "The number of casualties will be more than any of us can bear." It was a quote that would echo for years to come.

Giuliani was working under enormous personal strain. As mayor, he had always been exceedingly close to the police and firefighters. No matter what the hour, if a police officer was seriously injured or a firefighter killed in the line of duty, the mayor was among the first at the hospital, comforting the family members, visiting the wounded. If the worst occurred, he was always present for the funeral, and now it was becoming clear that the number of funerals would be unimaginable. The mayor also continued to get reports that friends and colleagues of his had perished. The husband of Giuliani's longtime personal secretary, Beth Petrone, was among the victims. Petrone's husband, Terry Hatton, was a Fire Department captain who commanded a rescue unit at the towers.

As the day unfolded, Giuliani shuttled back and forth between the police academy and Ground Zero, making five separate visits. The mayor also found time to visit Bellevue and St. Vincent's hospitals. "He is almost like God. People are coming up to him crying, thanking him for being there. All they want to do is make him say it's gonna be okay. And that's exactly what he does," recalled Bernard Kerik.[7]

AIR FORCE ONE landed at Offutt Air Force Base in Nebraska at 3:05 P.M. The president was whisked into an underground bunker, where he conducted a teleconference with his national security team. But a good deal of the president's attention was devoted to fighting with his aides about when he would return to Washington. The aides later said that plans were made to have the president address the nation from the Nebraska bunker, but Bush would have none of it. "At one point, he said he didn't want any tinhorn terrorist keeping him out of Washington," press secretary Ari Fleischer recalled. "That's verbatim."

Nevertheless, the president was persuaded to remain at the base until "the dust settles." Finally, Air Force One took off for Washington at 4:36 P.M.

At about the same time, Giuliani met with the city medical examiner. He was told for the first time that it was highly unlikely that any survivors would be recovered from the wreckage. As Chief King had intuited when he imagined he was about to die in the tower, the bodies of many of the victims had been vaporized. Nevertheless, the mayor ordered a round-the-clock rescue effort to begin at Ground Zero.

At 5:20 P.M., the 47-story office building at 7 World Trade Center collapsed, taking with it the city's vaunted command center. The mayor's day had become a blur of grief and horror punctuated by dozens of decisions that needed to be made and a series of public pronouncements. Yet he invariably struck the right tone.

To the public he mixed hope—"New York is still here. . . . We've suffered terrible losses and we will grieve for them, but we will be here, tomorrow and forever"—with a measured appeal for understanding: "Hatred, prejudice and anger is what caused this. . . . We should act bravely, we should act in a tolerant way." He promised that the city would survive. "We're going to get through it. It's just going to be a very, very difficult time. I—I don't think we yet know the pain we're going to feel when we find out who we've lost. But the thing we have to focus on now is getting the city through this, surviving and being stronger for it."

AT 8:30 P.M., President Bush finally delivered a five-minute address to the nation from the Oval Office. "These acts shattered steel, but they cannot dent the steel of American resolve," he said. The president then met again with his national security team and headed for bed, rejecting Secret Service pleas that he and his wife, Laura, remain in a secure bunker below the White House overnight. "Oh, no, we're not," Bush said. "I'm really tired. I've had a heck of a day and I'm going to sleep in my own bed." And he did, putting on his pajamas and getting into bed with his wife at 11 P.M.

ON THE EVENING of September 11, the living returned home. Virtually everyone who did not was dead, and the finality of it all would make the events of the day even harder to comprehend. There would be no second chances—no victims pulled from the brink of death by skilled medical treatment, even though it seemed half the doctors in the city were standing by at the scene, eager to be of help. The city and the nation were waiting for stories of rescues, and no rescues occurred. There were many people who had behaved bravely during the crisis. Anonymous white-collar workers helped disabled or overwhelmed colleagues down the stairs; some delayed their own evacuation to help total strangers who needed assistance. Teachers from schools in the shadow of the towers carried their students on their shoulders as they raced away from the danger. Due to the extraordinary performance of

Board of Education employees, every schoolchild who began the day in Lower Manhattan ended it in the arms of relieved parents.

Those simple acts of decency all seemed to merge into one general impression of New Yorkers rising to the occasion, no matter how dreadful the challenge. What the public seemed to be yearning for was the kind of happy endings Americans had come to expect from big disasters—civilians whisked away from certain doom by brave firefighters, or trapped victims unearthed by valiant rescue squads and returned to their relieved and joyful families. But almost nothing like that happened, and people were even more frightened and confused by the abruptness of what had occurred. Loved ones were simply and suddenly just—gone. Voices on cell phones vanished. Mike Pelletier's wife would never speak to her husband again. On the other hand, Eric Levine and thousands of others who were at the heart of the danger when the planes struck simply walked away from the worst disaster in the nation's history. The very thinness of the line between certain doom and the normal world would haunt the survivors long after any physical effects of their ordeal had vanished.

The scene at Stephen King's house in Brooklyn—a pale, worried wife opening the door, seeing her missing firefighter standing at the threshold, and breaking into sobs—happened for only a tragically few of the families who were waiting without word from husbands and sons who had raced toward Lower Manhattan earlier that day. "It was as if I had come back from the dead," King recalled. He sat on the living room couch, watching images of the towers collapsing on the television screen in front of him. King had never actually seen the towers come down. He began to shake uncontrollably.

The Fire Department had suffered an astonishing number of casualties on September 11—343 firefighters were killed at the Twin Towers. True to their code, they had run toward the disaster, and many were on their way up the stairs of the burning towers before the officers on the ground had any clear idea of who had gone where. "Our history has always been to go into burning buildings. It's ingrained," King said. "There was no way we were going to say, 'Wait. Let's think this out:

Yes, there's a couple of thousand people trapped in there, but if we go in we're just going to add a few more hundred to that number.' "

AT THE POLICE academy around 11 P.M., Giuliani issued orders for the next day and sent his staff home to get some rest. Then he returned to Ground Zero with Police Commissioner Kerik. "Going to Ground Zero that night is like going to hell," Kerik recalled. "I remember pulling up five blocks away. Everything is on fire. Rudy and I say nothing. There is nothing to say."

Giuliani watched shadowy rescue personnel working under floodlights dig through the smoldering rubble. Debris continued to tumble from what was left of the towers. The mayor closed his eyes several times, hoping to see the Twin Towers reappear when he opened them.

Around 2:30 A.M., he finally returned to the apartment of Howard Koeppel, the friend with whom the mayor had been staying since separating from his wife, who still lived in Gracie Mansion. Too tired even to shower, Giuliani flopped in front of a television set. Just as Stephen King was doing in Brooklyn, he stared at the screen and watched the towers disintegrate for the first time. He picked up a biography of Winston Churchill and read the chapters about Churchill becoming prime minister in 1940. He finally fell asleep about 4:30 A.M. He would be up and on the go in an hour.

THE SCENE AT Ground Zero—Firefighter King staggering, dazed, through a world of ash; Eric Levine trying to regain his footing as the South Tower began its ferocious disintegration; Rudy Giuliani dashing around under a rain of rubble, searching for a secure location from which to command—was all too real. But it was also a metaphor for the way the rest of the nation saw the world when it began again on September 12—as an eerie, unstable, confusing, and very foggy place with dangers so sinister they seemed to belong more properly to dreams or horror movies.

But Americans are by nature optimistic, and as they retold the story of September 11 over and over, they looked for a hero to put in the center, to find something positive in the trauma of the terrorist attack.

The disaster had been so complete that there were remarkably few candidates for the role. None of the people on the doomed airplanes had lived to tell which passengers or crew members had been the bravest, though the passengers who revolted on Flight 93 assumed a special role in the American memory. The world honored the extraordinary firefighters, police, and other emergency workers who gave their lives that day, but for the most part, they seemed like the heroes of old myths, plucked off the earth by inscrutable gods. They had simply vanished, along with the victims they were trying to rescue. Except for their families, they were remembered mainly en masse—the Bravest and the Finest and the Portraits of Grief. Their comrades who survived shunned the inevitable designation of "hero," sadly aware of how arbitrary their survival had been and how little they had actually been able to accomplish, despite all their efforts. George Bush's performance on 9/11 was hardly the stuff of legend.

It was Rudy Giuliani's story of quick response and personal fearlessness that provided a clean and reassuring narrative. When he stood up that day, covered in soot, he embodied the resolve of the nation. His name became the one Americans would instinctively connect to that date. A few months later, *Time* magazine would pick him as Person of the Year over Bush and the other finalist, Osama bin Laden.

IT WAS NOT the first time in Rudy Giuliani's career that he had stepped forward at a moment in time when people seemed to be in particular need of a heroic figure. An opera lover from his youth, the only child in a family that had almost given up hope of a child, he had grown up with an affinity for dramatic battles and center stage. He had been a federal prosecutor who focused on the biggest of targets—the top hierarchies in the mob, City Hall, and Wall Street. He had been elected mayor on a promise to restore order to a city that was reeling from the combined forces of crack-driven crime and red ink. It's possible, as his enemies have argued, that public safety and the economy were already on an upswing when he arrived, and that his law-and-order administration was simply in the right place at the right time. But if so, the timing was impeccable, and the image of firm stability he conveyed was perfectly attuned to the moment. His first term as mayor brought him national renown as a crime fighter who made New York livable again. The murder rate had plummeted, tourism revived, and economic prosperity made the city vibrant once again.

But the second term had been a slowly escalating disaster, and by September 11, the 57-year-old Giuliani appeared to have reached both the end of his career as mayor and a political dead end. His need to be the only hero in the story had driven away some of the strongest members of his team, and he was surrounded more and more by second-rate yes-men. The city, which admired his feisty stubbornness when the enemies were drug dealers and cop killers, had grown tired of a seemingly endless series of brawls and political catfights with schools chancellors, black neighborhoods, museums, rival politicians, and even hot dog vendors.

Searching for a new career path, Giuliani had let it be known that he was planning to run for the U.S. Senate in a race to replace Daniel Patrick Moynihan. While some people found it hard to imagine the mayor working happily as one of 100 egomaniacal equals, his supporters relished the idea of pitting Rudy, the lifelong New Yorker, against newcomer Hillary Clinton in a race the whole country would be watching. But the Giuliani campaign fizzled, never really taking off.

On April 27, 2000, the mayor announced that he had prostate cancer. There was a natural, and genuine, outpouring of sympathy from all around the city. Nevertheless, it was the beginning of the worst publicity of Giuliani's life. Less than a week after the cancer announcement, the *New York Post* published a photo of Giuliani with his "very good friend," Judith Nathan, and the mayor's long-troubled marriage began unraveling in the most public manner possible. He and his wife, Donna Hanover, cemented their separation in dueling press conferences, and the whole episode began to veer into parody when Giuliani's friends started leaking embarrassing details about the marital battle to the press. On May 19, 2000, the mayor dropped out of a U.S. Senate race he had never formally entered. "I've decided that what I should do is put my health first," Giuliani said. In September, barred by law from seeking a third term, he was running out the string as an unpopular lame-duck mayor.

What a difference a day made.

EARLY ON THE morning of September 12, 2001, a black car stopped in front of a Brooklyn apartment building, its headlights piercing the still-dark street. Michael Cohen, a psychologist, emerged onto the sidewalk and got in. Cohen was an author of an award-winning model for how to handle communications in the wake of a crisis, and he had been summoned to meet privately with Rudy Giuliani at the Police Training Academy in Manhattan. Cohen had been recommended to the mayor by Harold Levy, the school's chancellor—the psychologist had been working with the school system on unrelated issues. Contacted by Levy late on September 11, Cohen had been up all night trying to work out exactly what he should say. The last thing he wanted to do, Cohen told himself, was to waste anyone's time.

At the academy, Cohen and another psychologist, Paul Hoffman, sat down with Giuliani and Governor George Pataki. (Pataki would spend the entire post-9/11 period as the mayor's shadow, ever-present but seldom noticed.) They talked about the most important messages that the traumatized public needed to hear, and the mayor, Cohen remembered, took "copious notes, and he was very careful to be as exact and as actionable as possible." Among other things, Cohen explained, the mayor needed to be a "trusted voice"—speaking with the authority of a person who absolutely and totally believed whatever he was saying. The psychologist was pretty sure that Giuliani could handle that role. "The first thing I said to him was: 'Mayor Giuliani, I didn't vote for you and I've disagreed with just about every decision you've made within this last year, but I never for one moment didn't believe that you believed you were doing the right thing.' " Beyond that, Cohen said, the mayor had to be sure he was accurate. He could not tell the people they were safe, but he might be able to give them reasons to believe that the danger was subsiding. And third, he needed to give the public a "trauma story"—an account of what had happened that continued the story beyond the attack itself. "The trauma narrative cannot be: 'There are people who hate us, they took over some planes . . . people died,'" Cohen said. The story had to continue beyond the disaster. Not to a happy ending—that wasn't possible. But to a "positive response." And finally, he concluded, Giuliani needed to make people see that their own personal woes were part of a larger, collective story of tragedy, grief, and recovery.

They went from the meeting immediately into a press conference, and Cohen was surprised and pleased to see Giuliani already making use of his advice, instinctively molding the principles into the emergency at hand. "He was amazing to work with because he could see the flowers and he could see the trees and he could see the continent. It's an amazing mind. It was quite astonishing really," said Cohen, who would continue to serve as Giuliani's adviser throughout the rest of his mayoralty.

In many ways, Cohen was preaching to the choir. But the fact that Giuliani reached out for advice when there were so many pressing demands on his time suggested that the always self-confident mayor had had a moment of genuine humility as he assessed the demands of his new role. At any rate, in the days immediately after the disaster, Giuliani's performance appeared to be nearly flawless. He ran the city. He went to one firefighter's wake after another, promising that the city—and the Fire Department—would come back stronger than ever.

"I tell the children that nobody can take their daddy from them," Giuliani said in a televised interview. "That all these wonderful things that they're hearing about their dad is in them, that they're very special children. Because not everybody has a father that is a great man."

On September 14, the Gorumba family of Staten Island got a call from the mayor's office reminding them that Giuliani intended to walk bride-to-be Diane Gorumba down the aisle two days later. Diane's brother had been a firefighter, and when he died in action a year earlier, the family had lost its last male. That's when Giuliani promised to stand in his place at the wedding. Gail Gorumba, the mother of the bride, said Giuliani's representative told her that the mayor "wanted to be there now more than ever because he wanted to show everybody that the good times do go on. I was overwhelmed. Watching this man covered in dust and debris and—and—and trying to get the backbone of the city to stand up again and he's thinking of the Gorumba wedding." Telling her story on Oprah Winfrey's show a year later, Gail Gorumba said that when Giuliani lifted his daughter's veil, "not only did he kiss her on the cheek, he kissed the cheek of every New Yorker that day."

Leader and comforter, Giuliani united the city while helping it to imagine a better, stronger future. "We're not in a different world," he told *Time* in his Person of the Year profile. "It's the same world as before, except now we understand it better. The threat and danger were there, but now we recognize it. So it's probably a safer world now."

But the Person of the Year article went further than a celebration of Giuliani's valor on that critical day. Writing in the still-raw emotion of the postattack period, *Time* depicted him as an expert on terrorism who had foreseen the threat, prepared for it, and knew what needed to be done to avert a similar disaster in the future. It was a spin Giuliani endorsed. The idea that the mayor was a terrorism expert who knew what was coming became accepted wisdom—repeated by Giuliani's fans and by Giuliani himself. One example came in remarks he made to a crowd gathered for the 2002 screening of an HBO film on the World Trade Center attack featuring Giuliani and produced by his agent, Brad Grey. According to columnist Stanley Crouch in the *Daily News*, Rudy told the assembled press that "as mayor he had been obsessed with terrorism" since 1993, when terrorists bombed the World Trade Center for the first time, that he had "assumed there would be another attack" and had "held many meetings in which potential terrorist actions were discussed."

It would not have been surprising if Giuliani, with all his impressive federal law enforcement background, *had* been obsessed with terrorism. In addition to the trade center bombing, a few months later the FBI had raided a bomb factory in Queens, arresting Muslim extremists in the final stages of plans to detonate bombs all over the city. Indeed, years later, the 9/11 Commission deferred to Giuliani as if he was a counterterrorist pioneer, seeking his advice on organizing the FBI and preparing for future attacks. Giuliani's public testimony at commission hearings in May 2004 turned into a spectacle of such salutes. A few weeks before, he'd told the commission staff in a confidential interview: "In my first few years as mayor I thought there was a definite terrorist threat." He'd claimed that the 1993 WTC bombing convinced him there was a single terrorist network behind the ongoing threat. "A series of briefings that followed led me to believe that New York City was a target," he'd said, telling the committee that he was "advised" about tunnels, bridges, subways, "and other specific targets." It was a measure of how far ahead of the curve he depicted himself as being before the attack came.

No amount of praise seemed adequate. "No matter what happens now, Rudy Giuliani's legend is in place," wrote Jonathan Alter in *Newsweek*. "He's our Winston Churchill, walking the rubble, calming and inspiring his heartbroken but defiant people. His performance . . . is setting a new global standard for crisis leadership: strong, sensitive, straightforward and seriously well informed about every meaningful detail of the calamity. The mayor is unmistakably in command and every New Yorker—even those who dislike him—is gratefully in his debt."[8]

Time depicted Giuliani as a near-miracle of competence in the wake of the attacks: "In the hours that followed he had to lock parts of the city down and break others open, create a makeshift command center and a temporary morgue, find a million pairs of gloves and dust masks and respirators, throw up protections against another attack, tame the mobs that might go looking for vengeance and somehow persuade the rest of the city that it had not been fatally shot through the heart. It was up to Giuliani to hold off despair long enough for the rest of us to get our balance, find our armor and join in the fight at his side. . . . We knew that he was a tough man. It took the trauma for us to discover the tenderness, the offscreen, backstage, lowlight kindness he showed to widow after widow, child after child."[9]

Tim Russert, interviewing Giuliani on CNBC, announced: "Mr. Mayor, I'm going to take off my journalist's hat. I'm a citizen. I lived in the city for a long time. I worked at the World Trade Center. My son was born here. And on behalf of him and all the many others in this city and across the country, we want to thank you. We want to thank you for what you did in showing us the way through the extraordinary crisis."

Everyone, from heads of state to entertainment superstars, from the president of France to Oprah Winfrey to Muhammad Ali wanted to meet Rudy Giuliani and thank him for the way he had performed. And everyone wanted to be taken on a tour of Ground Zero by the mayor himself, past a police barricade that *New York Times* columnist John Tierney called the "the most exclusive velvet rope in town." When President Bush visited Lower Manhattan three days after the attack, it almost seemed as if the leader of the free world was there to grab some of the aura of command from the mayor of New York, rather than the other way around.

The events of September 11 restored Giuliani's political dreams—he had imagined himself as president since boyhood, but the personal and political disasters of 2000 had made any future in elective office seem virtually impossible. After 9/11, Giuliani became a Republican icon, at the top of most polls when the public was asked to pick the person they wanted the party to nominate in 2008. He was knighted in London and called to the witness stand in the death penalty phase of the Moussaoui case. When all levels of government proved humiliatingly inept in dealing with Hurricane Katrina, people said over and over again that what New Orleans needed was a Rudy Giuliani. The nation, which had attempted to define the kind of leader it required in the age of terrorism by choosing George Bush in the presidential election of 2004, thought again.

The new consensus candidate was obvious. Conservative Tom Roeser wrote in the *Chicago Sun-Times* that as he watched New Orleans going literally down the drain, "all I could think of was what Giuliani would do. Threats to our country demand a Prince of the City: a prince fearless enough to write his own rules to establish order."[10] When the president's hapless Federal Emergency Management Agency chief, Michael Brown, was hauled before Congress, Representative Chris Shays, a Republican from Connecticut, said wistfully that he couldn't help imagining "how different the answers would be if someone like Rudy Giuliani had been in your position instead of you."

"I never thought I'd sit here and be berated for not being Rudy Giuliani," Brown replied. "So I guess you want me to be the super-hero, to step in there and take everyone out of New Orleans."

CHAPTER 2

A DAY OF DYSFUNCTION

Giuliani understood the danger earlier than most. "I assumed from the time I came into office that New York City would be the subject of a terrorist attack," he says. The World Trade Center was bombed by Muslim terrorists in 1993, and while most New Yorkers pushed the memory aside, Giuliani did not. To ease the long-standing disaster-scene turf battles between police and fire, he created the Office of Emergency Management and built a $13 million emergency command center on the 23rd floor of 7 WTC, a mid-size building in the complex. The place was ridiculed as "Rudy's bunker." He and his staff held drills playing out 10 disaster scenarios, from anthrax attacks to truck bombs to poison-gas releases. He didn't foresee terrorists flying airliners into office towers, but the drilling ensured that when it happened, everyone in city government knew how to respond.

"PERSON OF THE YEAR: MAYOR OF THE WORLD,"
Time MAGAZINE, DECEMBER 31, 2001

THE MYTHMAKING ABOUT Rudy Giuliani began on September 11, and it involved an attempt to extrapolate his dramatic performance on that day into something more important. It also involved posttrauma fog—the focus on how the mayor had behaved in the moment of crisis, and not the emergency services his administration had put in place. It seemed unfeeling—almost *unpatriotic*—to ask whether Giuliani and his underlings had done enough to prepare New York for this terrible, although in some ways predictable, moment.

It was much less understandable, as the months wore on, to allow the afterglow of that moment to impose a national amnesia. Yet that was what happened. The facts—depressing but unavoidable—were that Giuliani had allowed the city to meet the disaster of September 11 unprepared in myriad ways. If the country had looked back dispassionately, it would have found that New York was left vulnerable on Giuliani's watch on several critical counts. His sine qua non of preparedness for disasters like this one, the Office of Emergency Management, had neither the power nor the capacity to cope with any grave crisis. The fire and police brass, which needed to work together, had not been taught—or if necessary forced—to do so. The command center had been built in the most vulnerable location possible, leaving the city without an operational headquarters when one was most needed. Giuliani had behaved from the outset of his mayoralty as if the 1993 bombing had never happened, ignoring its lessons and insulating himself from accessible intelligence about the continuing threat to his city. The public authority that owned the World Trade Center was encouraged to evade agreements binding it to comply with the building and fire codes. And the city's most critical agency that morning, the Fire Department, was allowed to retain ancient customs, faulty communications, and a culture that left it incapable of protecting itself, or others, in its worst moment of crisis.

This is the alternate version of 9/11. It focuses on the infrastructure of the day—the decisions and plans that caused the men and women who were actors in the crisis to do the things they did and to suffer the fates that befell them. The story begins with the agency that was supposed to be overseeing the response: the New York City Office of Emergency Management.

Mr. Kean: Mr. Mayor, just one thought. New York City on that terrible day in a sense was blessed because it had you as a leader. The commission is charged with making recommendations for the nation, and the rest of the cities in this country are not going to have a Mayor Giuliani.

They may have a good man or a good woman, but they're not going to have you. Have you got any thoughts about what kind of recommendations we could make, based on your experience, that would be across the board so that we could tell mayors of other cities—

Mr. Giuliani: I think the most important recommendation that I would make, put on the top of the list, is that cities should have Offices of Emergency Management. The Office of Emergency Management that we established in '95, '96, was invaluable to us. We would not have gotten through, when I say September 11, I don't just mean the day, I mean the months after that, and then the anthrax attack that followed it. Without OEM training us, doing drills, doing exercises, we would not have been able to handle all of that. If cities had that, it would help them a lot in terms of bringing together these resources, even in a city like New York as you found out.

PUBLIC HEARING, NATIONAL COMMISSION ON TERRORIST ATTACKS UPON THE UNITED STATES, MAY 19, 2004, CHAIRED BY THOMAS H. KEAN

Any attempt to establish a unified command on 9/11 would have been frustrated by the lack of communication and coordination among responding agencies. The Office of Emergency Management headquarters, which could have served as a focal point for information-sharing, was evacuated. Even prior to its evacuation, moreover, it did not play an integral role in ensuring that information was shared among agencies on 9/11.

STAFF STATEMENT #14, NATIONAL COMMISSION ON TERRORIST ATTACKS UPON THE UNITED STATES, RELEASED AT THE MAY 19, 2004, PUBLIC HEARING

NO ONE NOTICED the juxtaposition, at the same 2004 hearing, of Rudy Giuliani's endorsement of his own Office of Emergency Management as the indispensable national model of response and the 9/11 Commission staff's finding that it was invisible when the city's most deadly emergency required management. It was invisible before its command center at 7 World Trade Center was evacuated at 9:44 A.M., almost a full hour after the initial attack. And it was invisible in the tower lobbies and at the West Street incident command post nearby, where OEM's top officials, having rushed to the North Tower instead of their own command center, could find virtually nothing to do.

The city had actually had an office of emergency management, run by the Police Department, before Giuliani became mayor. Giuliani turned it into an independent agency directly under his control, and that was the only specific claim of programmatic foresight he could make from his pre-9/11 past. It became both the starting and the ending point whenever Giuliani champions tried to depict him as a prescient, antiterror warrior who, in the '90s, acted while others were distracted. Giuliani called OEM "one of the most important decisions I made" as mayor, in his best seller, *Leadership.* He even wrote that "as shocking" as the 9/11 attack was, "we had actually planned for just such a catastrophe," citing OEM's command center and drills as the sole evidence of such planning. It was also a prime theme of his many retrospective interviews—for example, he answered a 2003 question about what he did to secure the city after the 1993 attack by citing OEM and only OEM. "So then, on September 11, when it did happen in an unpredictable way," he said, "there was a lot more preparation for it than people would realize."

One of Giuliani's deputy mayors and longtime friends from his days as a prosecutor, Randy Mastro, appeared on CNN as the towers were burning on September 11, and was asked if the city had changed its approach since the 1993 attack. It sure had, he said, slamming the handling of the bombing by the Dinkins administration. "There was no coordinated city response," he said. "There was no Mayor's Office of Emergency Management. Rudy Giuliani established that. It's been one of the hallmarks of his tenure. And unfortunately, there are circumstances like this one where that coordinated effort has to come into play and it is coming into play right now."

But in fact, on the day when it was most needed, OEM was virtually useless.

Cate Taylor, a 9/11 Commission staffer who listened to all the radio transmissions that day, says simply, "There's nothing substantial on the OEM radios. OEM was not respected. Police and Fire don't see it as a coordinating agency." Sam Caspersen, one of the principal authors of the commission's final report on the city's response, concluded, "Nothing was happening at OEM" that had any "direct impact in the first 102 minutes of the rescue/evacuation operation."[1]

The Office of Emergency Management was directed by Richie Sheirer, a longtime city employee and Giuliani loyalist who had spent most of his career as a fire alarm dispatcher and a leader of one of the few unions to endorse Giuliani's 1993 candidacy, the dispatchers' union. Like many of the men closest to Giuliani, Sheirer's unspectacular career had suddenly turned meteoric when he attached himself to the mayor's coattails. A midlevel civil servant for decades, he was initially appointed top aide to Howard Safir, the head of the Fire Department, and then filled a similar position when Safir became police commissioner. From there, Sheirer leapfrogged into the top job at OEM, a role as a secondary 9/11 hero, and ultimately, a partner in Giuliani's postmayoralty consulting firm.

From the beginning, Sheirer's performance on 9/11 raised questions. Former New Jersey Attorney General John Farmer, who headed the 9/11 Commission unit that assessed the city response, found it strange that Sheirer, four OEM deputies, and a field responder went straight to the North Tower, where the Fire Department had set up a command post, rather than to the nearby emergency command center. "We tried to get a sense of what Sheirer was really doing. We tried to figure it out from the videos," Farmer said. "We couldn't tell. Everybody from OEM was with him, virtually the whole chain of command. Some of them should have been at the command center."

In many ways, OEM's dysfunction that day is the story of Sheirer's performance, partly because the rest of the unit's small top staff became an appendage to him. Jerry Hauer, Sheirer's predecessor, who is widely respected as an emergency management expert, said in a recent interview, "Richie should have early on pulled together an interagency command post. But he didn't command respect. He was lost. He was on his cell but he wasn't talking to the Police Department or Fire Department incident commanders. There was an enormous breakdown in the operations of OEM." Kevin Culley, a Fire Department captain who'd worked as a field responder at OEM virtually since its inception, was supposed to rush to the scene and did. Asked why most of the office's brass were also there, he said, "That's a good question. I don't know what they were doing. It was Sheirer's decision to go there on his own. The command center would normally be the focus of a major event and that would be where I would expect the director to be."

For whatever reason, Sheirer went straight to the North Tower. But once there, he and his team did not appear to be doing the most critical task assigned to them in a crisis—making sure various city agencies, particularly the fire and police departments, worked together and followed the protocols laid out for an emergency. It was a challenging mission, since the police and fire departments, if given their druthers, would often attempt to elbow each other out of the action. That appeared to be what happened in the North Tower lobby, under Sheirer's watch. The 9/11 Commission reported that a police rescue team "attempted to check-in" with the Fire Department chiefs in the North Tower at 9:15 A.M. and was "rebuffed," and that "the OEM personnel present did not intercede." A subsequent police team did not even attempt to check in, and no one at the North Tower enforced the protocols. Paul Browne, the current deputy police commissioner, said, "The Emergency Services Unit team is told to get lost and you can see Sheirer right there in the film footage from the day."[2]

The only example of a supposedly effective action at the North Tower that Sheirer cited later was his claim that he tried to set up a triage center in the lobby of 7 World Trade Center.[3] However, Richard Rotanz, who was Sheirer's top deputy still at 7 WTC, was never told about the triage order. Since the building had already been vacated as unsafe, establishing a triage area there made no sense and was never attempted, according to sources still in the tower at the time. But that wasn't Sheirer's only misfire that day. He recounted later that when Rotanz called from the command center and "told me there were other planes unaccounted for," Sheirer told the people in the lobby "that another plane was on the way," instantly converting unspecific information into a very specific false alarm. All Rotanz actually knew, from a Secret Service agent he saw in the building, was that there were unconfirmed reports of planes in the air. Sheirer's dangerous extrapolation spooked the chiefs in the lobby and quickly wound up on Fire and Police Department dispatches.

The camera in the North Tower caught much of what happened that morning, and at one point Sheirer's voice is heard in the lobby telling Fire Commissioner Tom Von Essen: "Tommy, Tommy, there's another plane coming."[4] Sheirer was so shaken by the threat of another plane that he "instructed the Police Department aviation unit to not let another plane hit." He told the commission that this bombast was an example of how "we were grasping at straws," since, obviously, no helicopter could "stop a commercial jet going over 400 miles per hour." Soon after Sheirer got the plane threat information, he left the lobby—because, he said, he was summoned to join Giuliani at 75 Barclay Street, where the police had set up a command post for the mayor.

Sheirer never went to OEM's command center on the day it was most needed. Nor did he go to the various active police or fire command posts on opposite sides of the towers, though OEM was supposed to make sure that the two departments worked together at a unified command post. Yet somehow, just a month after 9/11, Sheirer became a shining *New York* magazine profile, dubbed "the man behind the curtain" by Giuliani and "the wizard of OEM." In the article, Sheirer reeled off a list of theoretical 9/11 achievements, including placing calls for search and rescue teams—efforts that he attributed to others in his eventual commission testimony. He even claimed to have "helped move the Fire Department command post" out of the North Tower, though he never went to its new location on West Street in front of the World Financial Center.

Described by the magazine as "the short, stout man who bows his head in the limelight," Sheirer proved quite willing to celebrate his own modesty. "I think it's very valuable to be invisible for the job I do. We don't need fanfare," he said amid 5,270 words of nothing but.

"Invisibility enhances our ability to work with everyone because they know we're not looking to take the limelight."

It wasn't just the very visible post-9/11 Sheirer who was invisible on 9/11. His top deputy and eventual replacement, John Odermatt, was also at the North Tower and says that from the moment the first plane hit, he left only two staffers at the center. Commission investigators found that no one from any of the 60 public agencies or private entities like the hospital association and Con Ed that were assigned participants in OEM's emergency response ever showed up at the command center. (Odermatt says that "some came and were told to leave.") So the command center was out of business from the outset.

After nearly a year of investigation, the commission staff prepared a series of questions about OEM's shortcomings for Sheirer's 2004 public testimony.[5] Though the commissioners didn't ask most of these questions, and though many were clearly rhetorical, they resonate to this day. Noting that Sheirer stayed in the North Tower for an hour, the staff queried, "Why did you select this as your base of operations?" Pointing to "the large screen televisions and ability to monitor so many radio frequencies" at the command center, they asked, "was OEM conveying any information to incident managers who were struggling with limited situational awareness?" Observing that the Police Department, Fire Department, and Port Authority were "operating independently of each other" and not "coordinating their efforts" in the climb "towards the impact zones," the staff wondered "what steps OEM was taking that morning to address this?" No rationale for Sheirer's prolonged lobby stay, no information conveyed to commanders, and no steps to coordinate the response were ever discovered.

The agency was so irrelevant to the city's response that when Mayor Mike Bloomberg decided in 2002 to commission independent assessments, he asked McKinsey & Company, a major consulting firm, to do separate reports on the police and fire departments but didn't ask them to examine OEM's performance. When the McKinsey reports came out in 2002 and were critical of the city's response, it was Sheirer, a Giuliani Partner by then, who responded in the *Times*, blasting the report as "unprofessional" partly because McKinsey had not bothered to talk to him.

The *Times*' page-one, first-anniversary story by Al Baker offered its own implied rationale for McKinsey's supposed oversight, concluding that OEM "found itself marginalized and overwhelmed" on 9/11, adding that "the destruction of its command center was only the most visible of its problems." Baker's list of OEM failings that day—led by its inability to get the police and fire departments "to talk to one another"—was depicted as a consequence of an agency that "was plagued by politics from the start and later suffered a long period in which it languished." Calling OEM "perhaps publicly oversold," Baker contrasted the "trumpeting" of its "vision and effectiveness" by Giuliani with the fact that it had only held a single tabletop drill in the 18 months that Sheirer was in charge prior to 9/11. Even Odermatt, who limited his own public praise of the agency to its achievements in the weeks *after* 9/11, told Baker that there were things "that we could have done better" that day. When Odermatt became commissioner in the Bloomberg administration, he quickly rewrote most of the agency's emergency plans and named a deputy commissioner for preparedness, discovering a void 9/11 had made apparent.

Sheirer remained in charge of the agency for three months after Giuliani's term ended, and he departed without leaving in his wake a single meaningful piece of paper or disk about 9/11 or the lead-up to it. In the end, OEM was such a 9/11 nullity that even Giuliani, while praising it during his commission testimony, shifted midsentence to saluting its "invaluable" service "in the months after" the attack, a time-specific tribute emphasized by Odermatt, Caspersen, and others.

COMMAND AND CONTROL of police and fire response is critical in a massive emergency—so much so that in 2004, the Department of Homeland Security barred funding to cities that did not institute unified command practices that put one agency in charge of each anticipated emergency. Such practices would have been particularly critical in a city like New York, where the two departments had a longstanding and not-so-friendly rivalry. In *Leadership*, Giuliani described how he'd done just that, creating OEM to "make it clear that we could no longer have agencies individually responding to emergencies." The new "hybrid" crises, he wrote, required some overarching authority "that was equipped to coordinate many different departments. A bombing, for example, might result in an awful fire, but it's also a criminal act and may additionally require a response from other agencies. An attack of sarin gas would require intense coordination of the Health, Police and Fire Departments. I decided there should be an overarching agency—an Office of Emergency Management to coordinate and plan emergency response."

The *Times*' Baker was the first to reveal just how far Giuliani's description was from reality, and how undefined and disorganized the city's command and control was on 9/11. He called it an "undisputed, startling fact" that OEM had "failed to fully establish the most basic aspect of emergency response: determining who is in charge—and when and why." Baker wrote that the most fundamental part of OEM's mission—"having all agencies abide by a formal system that lays out under what circumstances police, fire or other officials take control of an emergency"—never existed in New York City "the way it does in other cities."

As axiomatic as the *Times* considered that assessment, Rudy Giuliani brazenly challenged it when he testified before the commission almost two years later. And just as no one noticed the bizarre juxtaposition of his OEM boasts at the 9/11 hearing and the commission staff's OEM critique, the contrast between his command and control claims and the staff's simultaneous findings of disarray also passed unnoticed.

Giuliani testified flatly that there "was not a problem of coordination on 9/11," while the staff explicitly found a "lack of coordination" and attributed it to police and fire command systems that were "designed to work independently, not together." Giuliani even dismissed decades of assumptions about the "war of the badges," insisting that the departments deferred to one another in "big emergencies," claiming he "saw several hundred of these and they've never proved to be a problem." If the two departments arrived at a scene and there "was any doubt" about who was in charge, Giuliani declared that the OEM director "had the authority to say Police Department in charge, Fire

Department in charge." That's why "you have to have a very strong OEM," he concluded.

Yet the day before Giuliani's testimony, the commission staff report had found that the two departments were "accustomed to responding independently to emergencies by 9/11" and that "neither had demonstrated the readiness to respond to an Incident Commander if that was an officer outside of their department." OEM, the staff concluded, "had not overcome this problem." Commission member John Lehman said the day the staff report was released that the city's "command and control and communications" were a scandal. "It's not worthy of the Boy Scouts, let alone this great city," the former Reagan administration navy secretary said, reflecting the staff's findings and provoking tabloid outrage. Lehman's statements, which occurred during his questioning of Sheirer, Von Essen, and Police Commissioner Bernie Kerik, weren't restricted to the performance that day, focusing as well on the period since the 1993 bombing. "It's a scandal that after laboring for eight years," he said, assailing a Giuliani directive on command and control issued in July 2001, "the city comes up with a plan for incident commanders that simply puts in concrete this clearly dysfunctional system."

The commission's final report echoed Lehman and the staff findings that "the FDNY and NYPD each considered itself operationally autonomous" and "as of September 11, were not prepared to comprehensively coordinate their efforts in responding to a major incident." In addition to the commission critique, Giuliani's expansive claims about OEM's ability to coordinate the prickly, independent forces were also contradicted by Sheirer's private testimony to commission investigators a month earlier, when he said that "OEM was not so much about being a referee" as it was "about assisting" fire and police. Sheirer added that it was OEM's "role to support the Incident Commander, whoever that may be," hardly sounding like the man who was supposed to pick one.

The starkest critique of OEM's performance came during the testimony of Edward Plaugher, the chief of the Arlington County Fire Department in Virginia, who led the emergency response to the Pentagon attack. The commission's report said that Plaugher and the other Pentagon responders "overcame the inherent complications of a response across jurisdictions because the Incident Command System—a formalized management structure for emergency response—was in place in the National Capital Region on 9/11." Plaugher delivered spellbinding testimony, contrasting the Pentagon and New York responses just an hour or so after Giuliani and his entourage left the hearing.

After Plaugher finished describing how Virginia, D.C., and Maryland worked together, Commissioner Jamie Gorelick said that Plaugher had told the staff "that the lack of a unified command" in New York "dramatically impacted the loss of first responder lives on 9/11." Asked if that was still his view, Plaugher deadpanned, "Yes, ma'am, it is."

"We, as a nation, have been practicing unified command for decades and we have honed this down to an absolute science," he explained, saluting the practice of putting one commander in charge of all the varied units responding to an event. "And so, for us now to be struggling with this is a little bit unconscionable for a lot of us in the profession. In March 2001, the fire chiefs and police chiefs from the entire Washington metropolitan region spent two full days at the National Fire Academy and worked out regional differences and left agreeing that unified command was how we were going to operate. And on 9/11, that's exactly what we did." Gorelick asked if "the ways things worked out in New York City" were "not consistent with the way unified commands work in the rest of the nation," and Plaugher said the city was "struggling with understanding the concept of who's in charge of what."

Remarkably, with Giuliani dominating the coverage, no one reported what Plaugher said. As telling as his public comments were, the 38-year fire service professional was even more outspoken when he met with the commission's investigators in the October 16, 2003 staff interview that Gorelick cited. He contended that "they built a system" in New York City "that allowed and encouraged that tragedy to occur."[6]

"If you truly build barriers," Plaugher observed, "you won't hear from the police helicopter that the buildings are going to fall," pointing out one of the deadliest consequences of New York's 9/11 segregated response (the police copter warned that "large pieces" would fall, but the fire department never heard the warning). Asked about his own 9/11 ride in a police helicopter over the Pentagon "in contrast to the lack of such assets provided the FDNY," Plaugher responded that "if you're locked into your own closed system, that type of thing is going to happen. You need to create a nurturing environment, not a closed one."

"Don't sugarcoat this," he told the staff. "When you think you're bigger than life—convince yourself of it—you're capable of doing things like the WTC response. They truly think they're bigger than life." The investigators noted: "Plaugher basically thinks that the loss of first responders in the 9/11 NYC response was of no great surprise to those within the community—'self-fulfilling prophecy.' "

Past president of the Virginia Fire Chiefs Association and a member of the Department of Homeland Security's Emergency Response Advisory Board, Plaugher said that New York's fire and police departments had "historically held the position that they are unique within the first responder community." The fire chiefs, he said, have consistently said that "it's impossible to run an incident in New York City with a unified command." That culture, he said, was a by-product of "the formative years of senior Fire Department leadership in the mid-70's and 1980's," when the department fought "thousands" of arson and other fires, and became convinced that they were "invincible." This "entrenched them into the 'go it alone' approach," concluded Plaugher.

The commission took a pass on Plaugher's tough tone, saying that "whether the lack of coordination" between the Fire Department and the Police Department had "a catastrophic effect"—meaning whether it killed people—was a "subject of controversy" with too many variables for the commission to "responsibly quantify those consequences." The final report did find, however, that the Incident Command System "did not function" in New York City. That was as flat and sweeping a conclusion as Giuliani's disingenuous denial. Sam Caspersen put it even more bluntly in an interview: "Anytime there isn't a fistfight between the Fire Department and the Police Department, they think it's an example of a coordinated response."

The National Institute of Standards and Technology issued its own 2005 report on the emergency response after a $23 million study that went into far more detail than the 9/11 Commission could. "An analysis of the overall city emergency response structure," the report found, "indicates that New York City did not have a formalized Incident Command System that encompassed citywide emergency response operations." Dismissing OEM as "basically a support organization for expediting emergency response," NIST concluded that "there was no formal structure of unified command between departments" and that interagency attempts to work together on 9/11 "were stymied by a lack of existing protocols that clearly defined authorities and responsibilities."

Bloomberg's McKinsey study reached a similar judgment: "Throughout the response on September 11, the FDNY and the NYPD rarely coordinated command and control functions and rarely exchanged information related to command and control." As unnoticed as these critiques were, they combined to destroy any argument that the Giuliani administration, through OEM, had achieved a prime objective as the mayor himself once described it: getting the emergency services on the same page.

COMMAND AND CONTROL failings were not the only city response errors with possible catastrophic effect that day. There was also the destruction of the ballyhooed Emergency Operations Center, which Giuliani had proudly positioned 23 flights up at 7 World Trade Center. The 9/11 Commission did not apply the same deadly calculus to the destruction of the command center as it tried to do to the costs of the command and control debacle. But the commission's senior counsel, John Farmer, cited a number of ways that an operating command center might have saved lives. If the center had been located elsewhere and thus able to remain operational that day, "I really think it would have made a difference. Maybe the failure to communicate among the agencies doesn't happen that day because that thing is functioning.

That's the point of it. I've never been convinced that they could have done that much better with civilians, but I think the number of responder deaths could have been greatly reduced. That's where I think the real tragedy is."

The National Institute of Standards and Technology and the McKinsey report also found that the loss of the command center made things worse, with NIST saying it "further impaired the coordination process." But neither spelled out the consequences of this loss like Farmer did, partly because the specifics are bizarrely both axiomatic and speculative. Examining the real-life costs of this lost command center is also a decidedly personal assignment of blame, since the siting was unquestionably a decision made by the hands-on mayor. Even Giuliani champions like *Time* called it "a mistake" to locate the command center on the 23rd floor of 7 WTC, part of a complex viewed long before 9/11 as one of the top terrorist targets in the world. Sheirer ridiculed it before and after 9/11. At a top-level meeting of the Giuliani administration, Police Commissioner Howard Safir fought the mayor's attempt to put it there, calling it "Ground Zero" years before 9/11 because of the 1993 attack.[7] And Lou Anemone, the highest-ranking uniformed police official, blasted it as well.

Giuliani, however, overruled all of this advice. Rejecting an already secure, technologically advanced city facility across the Brooklyn Bridge, he insisted on a command center within walking distance of City Hall, a curious standard quickly discarded by the Bloomberg administration, which instead put its center in Brooklyn. With then OEM head Jerry Hauer unwilling to put a facility underground in the City Hall area because of flooding concerns, Giuliani wound up settling in 1997 on the only bunker ever built in the clouds, at a site shaken to its foundation four years earlier by terrorists who vowed to return.

It was at once the dumbest decision he ever made and the one that made him a legend. If the center had been elsewhere, all the dramatic visuals that turned the soot-covered Giuliani into a nomad warrior would instead have been tense but tame footage from its barren press conference room, where reporters had been corralled prior to 9/11 for snowstorms and the millennium celebration. The closest cameras would've gotten to Giuliani-in-action would have been shots of him with a hundred officials from every conceivable agency and organization, working over monitors, maps, mikes, and phones. Had he been able to get into and operate from a command center he says he "headed" for shortly after 8:46 that morning, he might have been more effective, but he would also have been less inspirational.

Farmer says there's a legion of ways a functional command center might have helped spare lives that day. To start with, the towers were "configured in such a way that the fire chiefs told us they had no idea about the conditions on the upper floors," while the command center "would've had video to relay directly to the lobby." With every departmental radio frequency available and OEM, police, and fire brass with the mayor, Farmer says that "they would have had access to ongoing reports" from police helicopters, including that pilot's warning that the South Tower looked like it would partially collapse nine minutes before it did. The 9/11 Commission, the National Institute of Standards and Technology, and McKinsey all reported that this "potentially important information," as McKinsey put it, "never reached the Incident Commander," Pete Ganci, "or the senior FDNY chiefs in the lobbies."

Joe Pfeifer, the battalion chief who took initial command of the North Tower lobby and played a pivotal role all morning, put the pilot warning in the broader context of the overall information breakdown. "We weren't getting full intelligence reports of what was going on. I don't think you saw any ranking law enforcement in the lobby. The helicopters were up, but we had no means to communicate with them. We tried a number of times to do that, but it wasn't happening and no one was coming in to volunteer any information to us. Helicopters were able to assess the situation. My question is 'Who did they tell?' They didn't tell us. Or if they told anybody, whoever they told had a responsibility to tell us."[8]

Another vital cross-departmental source of information that might have been available to Ganci and Pfeifer had the command center been operative was the buildings department. Farmer says that "it might have been helpful to have someone at the command center that day who was a structural engineer" and, from videos, could have anticipated the collapse. Engineers from the Department of Buildings routinely appeared at the command center during emergencies and surely would have been summoned or sent there that morning.

The saga of John Peruggia, the OEM-assigned emergency medical chief is another example, like the pilot's unheard warning, of how a functioning command center might have affected lives. He was standing in the lobby of 7 WTC with a Department of Buildings engineer. Several minutes before the South Tower fell, the engineer warned them that he was "very confident" that the structural stability of both towers "had been compromised and that the North Tower, which was nearest to them, was 'in danger of imminent collapse.' " Peruggia sent a messenger by foot all the way to the West Street command post to convey the engineer's warning because he had no radio or phone that could reach Chief Ganci.[9] The messenger got to Ganci 30 seconds before the first collapse. The engineer made this judgment from the ground, and it was relayed literally on the ground. If a command center had been operating with engineers, videos, communications, and decision makers of every stripe, this judgment might have been made much earlier and communicated much more quickly.

That's why George Delgrosso, an ex–city detective who worked on Farmer's team at the commission, considers it obvious that a command center would have made a difference: "If there was a working center—anywhere—all information would have been going into one location."

The brass might even have found out about the horribly mismanaged 9-1-1, fire dispatch, and Port Authority police desk communications, all of which were telling civilians below the fires to stay in place long after full evacuations had been ordered by the chiefs. They also might have figured out a way to tell people above the fires not to head for the inaccessible roof, and to try instead to go down—a lifesaving option for the 18 people who were lucky enough to find the only open passageway, Stairway A in the South Tower. The clearance of occupied elevators and even the deployment of far more firefighters than required could well have been handled differently.

Farmer certainly thinks a functioning command center could also have tackled "the redundancy" of so many units, police and fire, checking the same floors again and again, lessening the need for all those firefighters on the stairs. Delgrosso put it succinctly: "If anybody kept a record of which floors were searched, they wouldn't have needed half those firefighters." The commission report concluded that "with a better understanding of the resources already available"—which certainly would have come with a functioning command center—"additional units might not have been dispatched to the South Tower at 9:37," a decision that led to fatalities. Farmer stressed in an interview that reducing these redundant and unnecessary forays was a prime way a command center could have saved lives.

Though Giuliani has been criticized for his siting decision, he's never been forced in any public appearance to deal with its possible consequences that day. He's made it clear that had the command center been functional, that's where he would have gone right away. In the fourth paragraph of his book, *Leadership*, right after describing how he got the news of the initial attack, Giuliani launched his best-selling leadership manual with a salute to the command center.

"My administration had built a state-of-the-art command center," he wrote, "from which we handled the emergencies that inevitably befall a city like New York, such as the West Nile virus, blackouts, heat waves, snowstorms and Y2K. It was packed with computers and television screens to monitor conditions all over the city and beyond. It had generators in case the power failed, sleeping accommodations in case we had to stay overnight, storage tanks with water and fuel, and stockpiles of various antidotes. So that's where Denny Young and I headed."

Despite this unambiguous statement, Giuliani has since quietly floated an alternative notion of where he was headed that morning. The only time he offered a detailed account of this other scenario was in his private testimony before the commission staff in 2004. "Even if the Emergency Operations Center had been available," he told the staff, "I would not have gone there for an hour or an hour and a half. I would want to spend some time at the actual incident, at operations command posts." He claimed he would have gone to the center "once everything was worked out on the streets." This revisionist version of events allowed him to duck the question of whether the location, and consequent loss, of the command center had life-or-death implications. But it's hard to see why Giuliani would insist that the command center be within walking distance of City Hall if he planned to first drive to the scene of the megadisaster, wherever it occurred. And many of the people who knew the situation best, including Jerry Hauer, his deputy Steven Kuhr, and Sheirer's first deputy, Odermatt, rejected the notion that the command center was a second stop, to be visited 90 minutes after an emergency. Kuhr said, "That's the kind of thing you hear from the uninitiated," and Odermatt insisted that "for sure" the command center would have been "the center of command" if it had been usable that morning.

In fact, the mayor's actions that morning spelled out both how clear it was that he intended to manage a crisis from his command facility and how pivotal a role a center could have played. Giuliani has acknowledged that when he got downtown and was told the command center was unsafe, "I immediately devised two priorities." The first was that "we had to set up a new command center." Leading an entourage of top aides that soon reached 25, he foraged for hours trying to find any semiuseful replacement, breaking into a locked firehouse at one point, and winding up at the Police Academy in midtown.

His top aides have made much the same point. Von Essen said that when he went to the command center and was told everyone was gone, he was really upset. "How can we be evacuating OEM now?" he remembered muttering. "What are we going to do, walk around all day?" He told a 2002 interviewer that he went to the command center because he "thought that was where we should all be" since that's what it was "built for."[10]

When Sheirer testified before the commission, he was as animated about the missing command center as Von Essen. Asked about the decision to put it at 7 WTC, Sheirer derided the location and blamed it on Hauer, as if the micromanaging mayor had had nothing to do with picking the site for a facility he'd planned in detail. In his private testimony, Sheirer tried to minimize what the loss meant, though he never argued that it was supposed to be used only after a situation was stabilized. And in his public appearance, Sheirer raved about its value. "That center was the most technologically advanced," he said. "It allowed us to deal with anything. The only problem is it had to be available to us when we really needed it. You need to make the commitment that this is a vital, vital facility. It should be in a hardened, security facility." Then, with his voice rising, looking up from behind his huge block glasses, he said, "Nothing should impede your ability to use it."

GIULIANI'S CLAIM IN a 2003 interview that creating OEM was "one of the first things that I did when I became mayor" missed the mark by nearly two and a half years. Even when the new agency started—27 months after Giuliani took office—it had almost no budget, only a director and a secretary on payroll, and borrowed space in a crammed city office. It stayed that way until the command center opened for operations, almost six and a half years after the 1993 bombing that Giuliani now says prompted it. The truth is that neither creating OEM nor building the command center had anything to do with the first bombing, even though the city's handling of that attack was a clarion call for just such a reorganization. OEM orchestrated 10 major drills and exercises by *Time* magazine's count, but none of them involved the targeted WTC complex or even replicated the 1993 attack elsewhere. After 9/11, Giuliani said he "assumed from the time" he was elected that the city would be attacked, a claim impossible to square with the years it took him to open his frontline center of defense.

It's one thing not to anticipate a plane wiping out multiple stories of a high-rise; it's another not to ever organize a multiagency, OEM-coordinated drill for what post-'93 fire experts started calling a "mega high-rise event" associated with a terrorist attack. According to OEM's Hauer, Steven Kuhr, and Kevin Culley, the agency never even had a tabletop exercise about a high-rise fire, either mega, terrorist-connected, or not. Maybe the high-rise fire scenario was all too familiar, and, as '93 and 9/11 established, all too hard to manage.

Even though a terrorist-provoked mega high-rise fire had already happened, and information culled from a much-publicized raid on a Queens bomb factory suggested it would happen again, high-rise preparation was treated as passe in the Giuliani era. The astonishing fact is that OEM never even developed a plan for a high-rise fire. Culley, whose job it was from 1997 to 9/11 to go to high-rise fires for OEM, said that "there was no OEM plan" other than "to ask the chief if there were any resources outside the Fire Department that he could use." He said OEM had "rough draft plans for minor emergencies" like a transit strike. "But I can't recall anybody anticipating another attack like the '93 bombing. We did not exercise any plan for something like that—there certainly was never any written plan for it." Because of both the '93 and the Oklahoma bombings, there was a concern about restricting parking by trucks and vans close to tall buildings, he says, "but no plan or exercise about physical terrorist attacks."

Kuhr, top deputy at OEM for four years, confirmed that the agency "never published a high-rise fire plan," and that "somewhere along the way someone probably decided that it wasn't necessary," since the Fire Department had one. "From a strategic perspective," Kuhr said, "I could understand why we would want to do a plan. It wouldn't have dealt with the tactics of firefighting; it would have been integrating the various agencies." But it was never seriously considered. And if OEM was relying on the Fire Department's plan, revised in Von Essen's second year, it was a very thin reed. As the 9/11 Commission's Caspersen pointed out, the plan "had no protocols for skyscraper collapse," much less any interagency authority.

New York firefighters had never been confronted with a blaze that consumed a full floor in a high-rise. Asked if the commission had found any evidence that the Fire Department had plans for how to fight such an enormous challenge, John Farmer responded that it had not. "Obviously, that's a huge problem," he added. "I don't know whose mistake it is. What is the strategy if the top ten floors are engulfed? How do you fight that? Everyone wishes that there should be a strategy for dealing with a catastrophe like that. What if it's the 20th floor that's engulfed? Is everyone going to die above that floor? There was no plan." As difficult as these questions are for every city with skyscrapers, the only city whose towers had already been bombed spent no systemic energy dissecting them, either at the Fire Department or at OEM.

OEM's original organization chart in 1996 had eight emergency management trainers, with only one of those listed as responsible for developing a "High Rise Fire Plan."[11] This trainer—a Fire Department captain assigned to OEM—had five other assignments, left fairly soon after he started, and never developed the plan. Steven Kuhr supervised him but doesn't recall anyone inheriting his fire plan responsibility. Half the trainers were assigned to work on a chemical/biological/nuclear response plan, which Kevin Culley said was "the focus" of the agency. OEM had "a plan to deliver pharmaceuticals," in case of a weapons-of-mass-destruction attack, Culley said, but not one to deliver occupants to the street from a fire that consumed one or more upper floors of a tower.

Referring to the list of blackout, storm, transit strike, and building collapse scenarios that OEM trained for, Senator Bob Kerrey, a member of the 9/11 Commission, asked Police Commissioner Bernie Kerik why "exercises dealing with the potential of al Qaeda attacking New York City one more time" were "notably absent from this list?" All Kerik could do was cite OEM's biological and chemical exercises. Giuliani also told the commission that the police and fire departments "could handle almost any emergency ad hoc, but what they were really behind on was biological and chemical attacks. I thought we needed an umbrella agency to train and push them" on that.

OEM ran detailed scenarios for WMD and Red X bubonic plague, but it was still counting on a largely unchanged Fire Department plan to handle the hot potato that had already left a '93 burn. It didn't seem to matter that no one in '93 had followed the FDNY's basic strategy—which was to tell occupants to remain where they were (called "defend-in-place") while firefighters ascended. Nor did it matter that the firefighters in '93 had taken nearly the same four hours to reach the top that it took the thousands of self-starting occupants to evacuate themselves.

If the usual problem is that generals obsess over the last war, the Giuliani team had the opposite inclination. It was fixated on the next one. As necessary as bio/chem preparation was, the primary terrorist threat predictably remained a high-rise fire, and the numerous ignored reports after 1993 about improving the city's response to that threat would, not just in hindsight, have made more useful mayoral reading than one more Churchill biography. But studies like the 1995 World Trade Center Task Force's examination of high-rise building code changes necessitated by the bombing—which was begun by Mayor David Dinkins's fire and buildings commissioners and completed by Giuliani's—were never acted on in City Hall or even turned over to OEM.[12] Hauer and Kuhr said that they had never heard of them. The after-action recommendations of the fire chiefs that oversaw the 1993 response also became dead letters, suggesting that six dead victims were not enough of a policy-changing mandate.[13]

Yet a high-rise fire was, of course, precisely what the city faced on 9/11. The best rebuttal of the Giuliani sound bite that no one could have anticipated what happened that day is Giuliani himself. He told Tim Russert in his first major television interview after September 11 that the city treated the attack as a fire and "we were basically following all the steps we normally follow in a rescue or fire." With that predictable a challenge, the calamitous response begs for explanation.

THE FIRE DEPARTMENT had the same radios in 2001 as it had in 1993. In fact, it essentially had been using the same radios since the 1960s, though they were more suitable for a burning three-story walk-up than for a steel inferno. The high-rise deficiencies of the "handie-talkie" radios were the most disastrous example of a failure to anticipate the needs of the emergency services in an age of terror.

The Giuliani administration waited until its final months in March 2001 to introduce a new digital radio that was potentially interoperable with police radios and better able to penetrate high-rises—new capacities that could finally correct the communications difficulties of 1993. The new radio, however, performed so badly that the Fire Department had to withdraw it within days of issuing it. Von Essen vowed to fix the quirks immediately and get the digitals back in the field, but six months later he and the mayor, each with a foot out the door, had done nothing. So, on 9/11, firefighters were still stuck with the old handie-talkies.

The only effort that made any headway in the eight Giuliani years to get police and fire personnel to talk to each other by radio was engineered by OEM's Jerry Hauer, who got interoperable 800-megahertz radios to chiefs in both departments. "Sitting on the shelf at fire dispatch," said Delgrosso, were the OEM radios that could have connected the two uniformed forces. "They took them out on weekends to check the batteries and put them back on the shelf." On 9/11, a day of such communications overload that everyone was stepping on everyone else's radioed messages, OEM's high-powered and interoperable radios were as deadly silent as ever. The Fire Department incident commander, Pete Ganci, was said by friends to have left his 800-megahertz radio in the trunk of his car.

In his prepared statement for the commission hearings, Hauer was almost penitent about the failing. "As I look back at September 11 and what might have had an impact on the number of people lost, I see our inability to get the departments to talk with one another on a common frequency as one of the issues that might have had an impact on reducing the loss of life," he said.

The 9/11 Commission found that the effectiveness of the handie-talkies was "drastically reduced in the high-rise environment," and that as many as eight Fire Department companies in the North Tower did not receive the evacuation order "via radio or directly from other first responders." It did not say how many additional companies got word of the evacuation only from "other first responders," but, obviously, firefighters are far less likely to respond to a secondhand evacuation order relayed by cops than to one they hear from a chief on their own radios. So more than eight companies experienced potentially deadly radio silence. While the commission attributed some of the communications difficulties to off-duty firefighters without radios, and some to firefighters who were tuned to the wrong channel, it did conclude that the radios themselves were a factor.

In the months immediately after 9/11, Von Essen and other Fire Department officials began blaming the communications breakdown on malfunctioning repeaters, or amplifiers, that the Port Authority installed after the '93 bombing. When the commission determined that the repeaters in each tower were functional that day, Giuliani tried another version of the same argument, contending that the chiefs in the lobby "decided" the repeater "wasn't operable" and that "they couldn't use it." Both positions appeared to be attempts to divert attention from the radios. But every investigation that's examined the 9/11 failings—including the commission, the National Institute of Standards and Technology, and McKinsey—refused to take that bait. These studies indicate that the collapse of the South Tower may well have damaged the North Tower repeater controls, as it did nearly everything else in the lobby next door. In addition, the chiefs fled the North Tower lobby right after the South Tower collapse, so there was no one left to operate the repeater. Consequently, the demise of the repeater after the 9:59 collapse means that the fire companies who never heard any of the subsequent North Tower evacuation orders—the critical message of the day—would have had to rely excessively on their radios, even if the repeater had initially been operative.

Beyond the dropped evacuation messages on the old radios, the three investigations also pointed to the dire consequences of the interoperability failures that day. When the commission said that "FDNY chiefs would have benefited greatly had they been able to communicate" with police helicopters, it was concerned not just about the missed pilot warning that the South Tower was about to fall. With interoperable radios, the chiefs would have also heard the pilots repeatedly warn of the North Tower collapse, as early as 25 minutes before it happened. In fact, some chiefs didn't even know that the South Tower had collapsed, though the pilots reported it immediately.

Nonetheless, the commission concluded that "the technical failure of FDNY radios, while a contributing factor, was not the primary cause of many firefighting fatalities in the North Tower." This murky language could mean that more firefighters died for reasons other than dysfunctional radios than those who died because of them, but since 343 of them died, that still leaves a long possible casualty list. It could also mean that the command and control breakdown, for example, had consequences more deadly than the radio malfunction, making it the "primary cause" of firefighter fatalities. In this choose-your-own-poison analysis of Giuliani preparedness disasters, radios could have even come in third, after the strategic miscalculation of flooding the stairwells with hundreds of firefighters when all they could do was rescue the immobile and trapped people who were below the fire. If the radios weren't the primary or even the secondary cause of fatalities, the commission nonetheless did find, almost despite itself, that they were a "contributing factor."

As with all the other response failings, the commission's 102-minute focused report did not comment on any of the radio history that preceded 9/11. But Sam Caspersen did. "The question of why nothing was done about the radios from 1993 to 1998," when the contract for the new digitals was awarded, "came up in multiple interviews and we never got a good response," he said later. "The question of why it took three years to get radios into the hands of firefighters in 2001 did not come up precisely because our number one job was to tell what happened that day."

In any event, the National Institute of Standards and Technology reached a more straightforward conclusion than the commission in 2005. "A preponderance of the evidence indicates that emergency responder lives were likely lost at the WTC resulting from the lack of timely information-sharing and inadequate communications capabilities," the report stated, with none of the commission's fuzziness. It said only half of responders "heard radio messages calling for the immediate evacuation of the building." Usually a master of understatement, NIST approvingly cited a firefighter who said simply, "If communications were better, more firefighters might have lived."

MOST CALLS FROM World Trade Center occupants that morning were transferred by 9-1-1 operators to fire dispatch, and no unit in the department was more institutionally isolated than the one that interfaced with fire victims. Caspersen found it "very upsetting to listen to phone calls and hear callers transferred to fire dispatch and get put on hold or get lost altogether." He also found it upsetting that there was "no change" during those 102 minutes in the "stay-put" advice of 9-1-1 operators and fire dispatchers "before the collapse of both towers." The *Times* discovered that only 2 of the 130 callers whose taped conversations they reviewed were told to leave. Though the fire chiefs had ordered a full evacuation from the start, the operators and dispatchers, who worked for the police and fire departments respectively, were still reading from the defend-in-place script, whose stay-put premise was disparaged in every 1993 bombing critique.

If the callers did not get useful directions through the operators from the fire chiefs on the scene, it's equally distressing that the chiefs didn't hear what the telephone operators were learning about conditions inside the building. Caspersen concluded that "no first responder group"—police or fire—"had the dedicated job of aggregating all of this information" that came in from 9-1-1 and dispatch and "disseminating it to chiefs at the emergency scene" to maximize their understanding of what was going on in the buildings. An internal commission document called it "a ruptured flow of information" in both directions. It particularly lamented that no one had anticipated "that extraordinary circumstances could arise where standard operating procedures"—like stay-where-you-are scripts—"could actually translate to harmful advice." Philip Zelikow, the commission's cautious executive director, said the operators "could have saved a number of additional lives if they had known what the people on the scene knew," meaning the chiefs.[14]

The identifiable casualties start below the fire—on the 64th and 83rd floors of the North Tower. Eileen Hooey, the wife of a Port Authority executive who died, says that the transcripts "clearly show" that her husband and his coworkers remained on the 64th floor because they "were instructed by both 9-1-1 and the Authority to stay in place." The commission confirmed that 16 other civilian employees of the Port Authority who could have left at any time instead sealed the doors to their office with tape and waited until 10:12 to leave, having been told repeatedly to remain there. Fourteen died as they finally descended, with two surviving a fall into a void. On the 83rd floor, 13 died after repeatedly asking 9-1-1 operators if the fire was above or below them. This group, trapped by debris, was transferred back and forth several times by dispatchers and told to stay put. It's unclear that they could have made it out, but the instructions to wait were clearly counterproductive.

But the "disconnection in the 9-1-1 loop," as John Farmer calls it, had potentially its greatest impact above the fire in the South Tower. The roof was locked, and the only way out was the "debris-cluttered" Stairway A, which did become the path to life for 18 people on the higher floors. The operators told caller after caller from the 25 floors above the fire to stay put. Had they told them to search for an open stairwell, more might have found it. Callers told them a few hundred people were huddled together in South Tower offices, unaware of the order to evacuate because the operators they called were also unaware of it. Asked what to do, one operator said, "I wish I could tell you. I don't know anything more than what people calling in tell me. I don't have any access to any radio or TV or anything." Another said to a fellow operator, "Are they still standing? The World Trade Center is there, right?" As uninformed as they were, the operators remained emphatic right to the end about what the callers should do. When one man just four floors above the fire indicated he was going to try to leave, the dispatcher insisted, "You cannot. You have to wait until somebody comes there."

The operators also didn't know that the roof doors were locked, so when Sean Rooney, who worked for Aon on the 98th floor, kept calling, he was never told it was a dead end. Rooney worked his way to the roof, pounding helplessly on the locked door, frustrated by his memories of the 1993 rooftop rescues by police helicopters. "I know this Stairway A was clear," Rooney's wife, Beverly Eckert, says. "I just don't know what prevented him from reaching it. They thought they had the option of going out on the roof. No signs said 'no rooftop egress.' In his mind, it's like '93; we will be rescued. They never made the place evacuable. I'm in the insurance business and I came to learn that they prevented landings on the roof. They never told anybody that option was foreclosed. If Sean found that one staircase was impassable, he and the others would have kept searching. The belief that they had a rooftop option cost them their lives.

"It took me a while to sort through what Sean was saying. He was mad at the Port Authority and the terrorists. He knew the terrorists had done this. He had colorful things to say about them, and the Port Authority. I heard the explosion and I heard him falling as the floor dropped out from under him. I called his name. The tower was collapsing as I turned to the TV to see with my eyes what I already knew in my heart."

There are even indications that both the Port Authority and the Fire Department knew about Stairway A. John Mongello, who worked at the authority's operations desk, said in a recorded transmission that a stairwell was free from the 98th floor down. Ladder 15 firefighter Scott Larsen radioed his lieutenant: "Just got a report from the director of Morgan Stanley. They say the stairway is clear all the way up."[15] There was certainly a way for responders to find out. "It's so easy," says John Farmer, "to have someone standing at the bottom of the stairwell asking civilians what floor they came from and what conditions are like. I believe they did that in 1993. That's something that could have been an easy thing." Even Rudy Giuliani later indicated that a fire chief told him there was a passable stairway, though he thought it was in the North Tower. "There were people on the 104th floor who walked all the way down," Giuliani said later, "if they were lucky enough to find the stairs."

"This lack of information, combined with the general advice to remain where they were," the 9/11 Commission report said, "may have caused civilians above the impact not to descend, although Stairway A may have been passable." In fact, an emergency service cop on his way up the South Tower steps just before it collapsed radioed that he'd just seen "a stream of civilians" coming down. Neither the cop nor the civilians survived. Since only 11 of those who died in the South Tower were occupants of below-the-fire floors, this large group obviously came from above the fire. They must have found Stairway A, but only shortly before the collapse, restrained perhaps by the stay-put directions of the 9-1-1 operators.

The death toll above the fire in the South Tower, while half the North Tower number, is twice as frustrating to the families—619 what-ifs. If they'd been told about the open stairway. If they'd been warned that the roof was in lockdown. Even if they'd simply been warned not to wait, but to search for a way down.

While many of the challenges presented by a megafire are daunting, creating an effective information stream to and from dispatchers isn't. Yet there's no indication that Von Essen, any of the three Giuliani police commissioners, or the mayor ever so much as considered how to create an informed loop between 9-1-1 operators, fire dispatchers, and on-scene commanders prior to 9/11. During the Giuliani years, the city collected over $250 million in telephone surcharges to upgrade 9-1-1 but diverted the majority of that funding for other budgetary purposes, even in years of huge surpluses. None of the three major improvements that were supposed to be funded by this surcharge, including computer-assisted dispatching, ever occurred, and some were as much as six years behind schedule. Giuliani resisted all efforts, some of which originated within his own administration, to take this mismanaged operation and planned upgrade away from the Police Department, reassigning it to the agency that oversaw information and technological systems. Unifying 9-1-1, Emergency Medical Services, and fire dispatch, as is done in other major cities, was never considered. Instead, Giuliani just let it drift.

One of the forces against any dramatic organizational or personnel change was the Fire Alarm Dispatchers, a small union that endorsed Giuliani in 1993 and 1997. The dispatchers provided security and drivers for the 1993 campaign, according to union president David Rozenzweig. One reason for the union's clout in Giuliani circles was its long-standing ties to Sheirer. "I'm sure he was involved in the campaign and the endorsement," Rozenzweig recalled. Sheirer got so close to the Giuliani camp during the campaign that he was asked to play a key role in orchestrating the inaugural, and was immediately promoted at the Fire Department.[16]

Yet Sheirer, with 22 years of fire dispatch experience, did nothing to close the loop when he rose to the top of the FDNY in the Giuliani era, nor did he when he moved on to the Police Department or the helm of emergency management in the later Giuliani years. As crucial as these emergency communications are—the only connection between the public and public safety—none of OEM's celebrated drills, before or after Sheirer, rigorously tested the system. Treated like civilian sideshows inside uniformed agencies, the 9-1-1 and dispatch units became unexamined wastelands.

AS MISMANAGED AS the dispatch communications with the victims at the top of the towers were, what was really startling in the aftermath reviews of the response was how badly botched the rescue operation was as well on the ground floor. Chris Young, a 33-year-old temporary worker, was the solitary occupant of an elevator stuck in a North Tower lobby flooded with firefighters.[17] He talked with Port Authority staff via intercom twice, and they were passing lists of occupied elevators to the fire command post. Yet no one came. He and the seven people trapped in an adjacent elevator could hear each other and worked out a deal that the first one out would get help for the others. Both Young and the other group kept trying to push and squeeze their way out. When Judith Martin and her companions forced open their double doors after 44 minutes, they walked through a sea of uniforms and told fire officers all about Chris Young. "The rescuers nodded okay, okay," one of them recalled.

Twenty minutes later, a violent explosion shook Young's elevator and threw him to the floor. The South Tower had collapsed. The destruction virtually emptied the North Tower lobby of firefighters, but it also disabled the motors that had kept Young's elevator doors shut. Young tried pushing again, and this time he got out, made his way through a broken window, and, five minutes before the North Tower crashed, evacuated himself. *USA Today* found that at least 21 people had done the same. It also found that 200 people had died in elevators, many simply trapped. The paper quoted the Port Authority manager who coordinated the intercom calls into the elevators. "A couple of firemen grabbed a couple of different lists," the manager said, "but I don't know what happened after that."

During the 1993 rescue operation, 500 people were trapped in elevators for hours. Elevator mechanics working for the Port Authority began manually releasing the elevators stuck in shafts, sending them to the nearest floor and popping doors open. But the 83 mechanics evacuated after the second plane hit on 9/11, fleeing to locations near the complex. The maintenance supervisor for the mechanics, James O'Neill, says no one asked them to return until an hour after the initial attack, and that he and another supervisor were on their way back to the South Tower when it collapsed. The mechanics also told *Elevator World*, the industry bible, that they "were instructed by city emergency personnel to get out of the building and safely away from the complex" when they evacuated in the first place. Without the mechanics, the Fire Department was lost.

It had taken four years after the 1993 debacle—when 72 schoolchildren were trapped for hours—for the FDNY to issue a new set of elevator instructions. But the 30-page training bulletin suggested no training. It said the function of the department "is limited to the safe removal of persons trapped in the elevator car" and that the "reactivation of elevators was not to be carried out by firefighters." As reasonable as that division of labor might appear, it's often impossible to remove anyone from an elevator unless the car is reactivated and brought to a landing. Firefighters were instructed to hit the "door open" button and the lobby call button. After that, they were told to "summon an elevator mechanic."

The 9/11 Commission's Farmer said, "There was no systematic department approach to assisting people in getting out of an elevator." Caspersen concurred, saying he didn't believe a Fire Department protocol existed for checking even lobby elevators. "They could have done more to ascertain if people were in the elevators," said Caspersen. "More creative leaders would have had firefighters checking lobby elevators." Firefighters are equipped with the Jaws of Life, a tool designed to force open elevator doors, but no one in the North Tower used one to free Chris Young or Judith Martin. A more systematic effort did occur in the South Tower, where firefighters had far less time and had begun using the Jaws to try to rescue a group of people in one elevator, when the tower collapsed, killing everyone.

Legendary fire commissioner John O'Hagan, still the guru of high-rise firefighting three decades after his textbook on it appeared, wrote that elevators had to be immediately "located, entered, and searched even when the fire forces are stretched thin." Any "compromise in this commitment can result in a loss of life," he wrote. But when faced with the most urgent need to act on that advice, eight years after a call for strategic change at the same location, the department had no viable plan of action.

"COURAGE ALONE, HOWEVER inspiring, is not always enough," wrote Tom Von Essen in the final pages of his book. The commissioner's words were a coda for the deficiencies in city leadership that almost certainly cost the lives of firefighters on 9/11. In contrast to "New York's Finest," the paramilitary Police Department with its emphasis on top-down command, the Fire Department revolved around individual firehouses. Firefighters drew their job satisfaction from their relationships with the other men—and it was almost always men—with whom they lived and worked. The culture of the firehouse had been reaffirmed in the 1970s and early 1980s, when arson became endemic in the city's worst neighborhoods. The firefighters, sometimes stranded in desolate areas of abandoned buildings, racing from one tenement blaze to another and occasionally assaulted by angry mobs eager to take out their antiauthoritarianism on any uniform in sight, had good reason to believe they could count on nothing but their own primal code of honor—to rush into the fire, to save victims, never to abandon a fellow firefighter.

It was a noble legacy, but in an age of skyscrapers and terrorism, it needed to be reined in by a strong leadership that understood that in a large-scale crisis, everyone needed to be coordinated and directed from above. That never happened. "The department has deep cultural roots in the power structure of New York," said John Lehman, the 9/11 commissioner who focused the most on frontline defense issues. "Rudy Giuliani was not about to impose himself on that structure. This was not just a failing of Rudy; every mayor has made that accommodation."[18] But Giuliani, who had a particularly close relationship with the department, was better positioned both to understand the problems and to enforce a solution. The fact that he never did had tragic consequences when the World Trade Center was attacked.

Giuliani's first fire commissioner, Howard Safir, was a longtime associate from the mayor's Justice Department days. Von Essen later described Safir as "a total outsider who knew nothing about our department." Safir's entire career was in law enforcement. He even continued to run a private security-consulting firm out of his Virginia house, theoretically on his own time, while serving as New York's fire commissioner. Everyone understood he was in line to replace Police Commissioner Bill Bratton, who had no long-term ties to Giuliani. One deputy mayor described Safir as "police commissioner-in-waiting," and in late April 1996, he did indeed succeed Bratton.

Safir, who did as little to rock the boat at the Fire Department as possible, was replaced by Von Essen, the first union leader in city history to become the commissioner of his own department. It hardly mattered that Von Essen had flunked the lieutenant's exam twice and never moved up the promotional ranks a single step. He had accomplished something else of transcendent importance: he delivered his union to Rudy Giuliani in the 1993 election. As rare as it was for a city union to break with an incumbent, Von Essen's ran Giuliani's biggest phone banks and dispatched an election-day army of firefighters to polling sites all over the city. "I was mingling at Giuliani's campaign party," Von Essen wrote in his book, "and standing not far from him when it was announced that he'd eked out a close victory. When I offered him my congratulations, he hugged me and said, 'You guys are unbelievable. They were everywhere.' "

Unlike Safir, Von Essen understood the department well and he did have an agenda for change: he wanted to take on the power of the fire chiefs, who were represented by a union that backed Dinkins, the Uniformed Fire Officers Association. Von Essen pointed out—with some accuracy—that while the chiefs were a critical part of the department management structure, they were protected by tenure, promoted by test, untrained as managers, and accountable to no one. He derided his own department in his private 9/11 Commission testimony, comparing it unfavorably to other agencies. "OEM was a meritocracy," he said, "unlike the FDNY."

In a foreword for Von Essen's book, Giuliani wrote that "Tom reformed a department that had remained largely unchanged since its inception." Yet Von Essen's solitary claim to reform in the 288-page memoir was that he reassigned a dozen chiefs from administrative to field duties, so small a revolution that even Giuliani couldn't turn it into a sound bite. Giuliani also praised Von Essen for "record low civilian casualties," but that was hardly a product of Fire Department reform, given the fact that the number of fires was down 46 percent.

Von Essen was in many ways one of the most introspective of Giuliani's team, and his post-9/11 assessments of the department's performance were stark and intelligent. But he was also commissioner for five and a half years and did nothing to put together a real plan to handle high-rise fires except issue a new circular that changed little. Though he always maintained that firefighter safety was his top issue, he left intact a safety battalion without any structural engineers, manned by fire officers who learned how to evaluate potential collapses by riding around with a veteran chief and watching him do it.[19] His department also did nothing to support Fire Marshal Ronnie Bucca's efforts to become the sole FDNY representative on the FBI-run Joint Terrorism Task Force—attempts that were rebuffed by the police, even though Bucca had a top security clearance and an encyclopedic knowledge of Islamic fundamentalists.[20] Asked about access to threat information by commission investigators, Von Essen, whose department had every reason to be included since it led the response to the 1993 bombing, said he was "told nothing at all." At the end of the long day of the terror attacks, Von Essen said he started hearing "talk of an organization called al Qaeda and a man named Osama bin Laden." But, he wrote, "it meant nothing to me."

And Giuliani, the one-obsession-at-a-time mayor, spent virtually no time worrying about the department's ossified command and planning structure. The index of his memoir contains 43 pages of pre-9/11 references to the Police Department versus 1 page for the Fire Department, noting merely that he met weekly with the fire commissioner. He did create OEM, which was both a by-product of and a boon to the Fire Department, because it was a check on police emergency power. But his focus was on crime fighting, where he was making a national reputation for bringing down the felony rate. There was no way for Giuliani to believably lay claim to the nationally plummeting fire fatality numbers. So he let the Fire Department become a sleepy backwater of patronage and paralysis.

NO FIRE CHIEF was more celebrated after 9/11—in moving *Newsweek* and *New Yorker* portraits—than the legendary Bill Feehan, a four-decade veteran who'd held every position in the department, including commissioner. A John Wayne style protagonist who would not even use a touch-tone telephone, Feehan believed it was important for soldiers and firemen "not to know too much." They had to "act on blind faith" without measuring the odds.

Feehan raced to the attack scene on 9/11 in a car of chiefs, watching the fire in stoic silence. They did not talk strategy; they knew feet and faith would drive the response, like they always had. The minute they pulled up to the towers, Feehan's driver started telling people to "get out of here" because the tower was "coming down." It was a prophetic warning, and hardly the only one from the FDNY. Yet, an hour later, when the South Tower did fall, the doughty Feehan turned to his stunned fellow firefighters and asked, "You expect this is going to happen twice?" He was standing with special operations chief Ray Downey, who'd predicted a collapse just like Feehan's driver. Belatedly convinced that their command post, on West Street opposite the towers, was vulnerable, Feehan, Downey and Pete Ganci, the top uniformed leaders of the department, ordered it moved north, farther away from the only tower left. Ganci also began issuing radio commands for firefighters in the North Tower to evacuate at 10:15, repeating it five times. But then, the three chiefs, a hundred years of combined experience, went south themselves, closer to the remaining tower.

"I'm going to take a walk down there," Ganci was heard saying as they charged into the dust. The three men picking their way through the rubble embodied the stout fearlessness of the department—as well as its lack of a plan that went much beyond walking directly toward danger. For one thing, many firefighters weren't abiding by the central assumption that Ganci and Feehan had determined would guide the response from the very beginning: that it was impossible to put out the fire or to save anyone on the floors above the blaze. The *Fire Officer's Handbook of Tactics* says that "even a light load fire" in a 30,000-square-foot floor—and this was five 40,000-square-foot floors—"is beyond the scope of manual firefighting." If the wishes of the top chiefs had been communicated down the line or followed by those who heard them, firefighters would have focused solely on rescuing people who were stuck on the floors below. Yet lower-ranking officers were sending their men up into the towers—and to their deaths—with orders to extinguish a blaze that their superiors had deemed utterly uncontrollable.

The absence of a real plan tailored to megafires—along with the predicted communications chasm—were precursors for the disparate response. Some of the department's missteps were understandable because of the shocking magnitude of the attack; others were an indictment of its readiness and competence. The commission's final assessment couldn't have been worse: "Certainly, the FDNY was not 'responsible for the management of the City's response to the emergency,' as the Mayor's directive would have required." Significant shortcomings within the department's command and control capabilities "were painfully exposed," said the commission.

Nicholas Scoppetta, the Bloomberg-appointed fire commissioner and longtime Giuliani friend, testified at a 2002 City Council hearing that "there would be a dramatic difference" in how the department would handle a "similar occurrence" today. "That would be certainly in deployment of resources, deployment of personnel, staging of personnel, and I think the entire strategic approach to an incident like that would be different," he said.

After the 1993 bombing, for instance, the department was well aware that one big evacuation problem involved the rescue of handicapped, severely overweight, or otherwise immobile people. It was a major issue that arose in fire journals, after-action reports, and even the World Trade Center Task Force report issued in 1995 by Giuliani's buildings and fire commissioners. But nothing was done to formalize a better plan, and when the WTC was attacked again the rescuers had nothing to guide them but their own instincts. So some firefighters diverted people who were slowing down the evacuation of others to waiting areas. As many as 40 to 60 people were sequestered between the 12th and 20th floors of the North Tower. Some stayed there until they died.[21]

The stockpiling of the impaired on "holding floors" appeared inconsistent with the prime purpose of the department's ascent, which, as Ganci and the chiefs determined it, was to rescue those who couldn't get themselves out. Since such sequestering never happened in the South Tower, it was also an impromptu decision, implemented in the glaring absence of so much as a single sentence about the disabled in the high-rise manual issued by the department in 1997. Many of the impaired were actually kept on the same floors where large numbers of ascending firefighters were resting.

Meanwhile, some evacuees were helping them make their way downstairs. This was particularly manageable with lightweight evacuation chairs that slid down the stairs, but the firefighters apparently had no idea of where they could be found. The only people known to have been carried down on these chairs were Port Authority employees.[22]

There was no single story on 9/11 more tragic than that of paraplegic Ed Beyea and his friend Abe Zelmanowitz. The two Blue Cross computer analysts waited on the 27th floor for nearly all the 102 minutes, seeking help to carry Beyea's heavy, motorized wheelchair down the stairs. Zelmanowitz told his brother Jack on the phone shortly after the plane hit, "We're just waiting for assistance. Firemen will come and help us down." They did come—streaming past them on the way up. An elevator was working up to the 24th floor for some of the time the two men were waiting—Beyea unable to move, Zelmanowitz rejecting all suggestions that he leave his friend behind. In the end, the two men died in the tower, along with Captain William Burke, who sent his men down and stayed with Beyea and Zelmanowitz in a last-minute effort to get them out. Burke's sister, Janet Roy, who pieced together a chronology from the surviving members of his unit, says they were "trying to figure out where the specific chair you use for evacuation was."

Another captain whose unit was on the 27th floor, Jay Jonas, rushed down after the South Tower collapse, having seen it from the window with Burke. Jonas's company stopped their own descent to pick up Josephine Harris, who was also impaired, and the delay in carrying her down almost cost those in Jonas's company their lives. They only made it to the fourth floor but survived in a free fall.

There was neither a standard procedure nor an order that morning making it clear what companies were to do with the impaired occupants they came across. Burke's unit never stopped in the lobby for any sort of orders. Captain Jay Jonas's unit did, and he was told: "Just go upstairs and do the best you can."[23]

THE SCORECARD OF Fire Department errors that jeopardized its own brave men—without considering civilian lives—is a challenge for change. The agony of the loss has for too long silenced the litany. Even when the investigators from the commission, the National Institute of Standards and Technology, and McKinsey uncovered terrible truths, they often shrank from the tough language that might have jarred a city, a nation, and especially a fire department, to greater reform. With the Victim Compensation Fund offering millions of dollars and a form of closure, all of the responder and civilian families who accepted the compensation forfeited their right to sue the city. The handful that didn't take the compensation focused their lawsuits on the airlines and others, not filing against the city. Because of these very individual decisions, no one in command from the city—from Giuliani to Sheirer to Von Essen—will ever be deposed about the city's fatal failings in the lead-up to attack or on that day. The only lawsuits against the city that remain revolve around post-9/11 toxic exposure and respiratory damage. Yet the widows and children of the 421 responders who died have to live with the question of whether their loss was due to municipal malfunction as well as distant madmen.

In Giuliani's appearance at the commission's 2004 hearing, he tried to make any discussion of what he conceded were "mistakes" unpatriotic, insisting that "our anger should clearly be directed, and the blame should clearly be directed at one source and one source alone, the terrorists who killed our loved ones." Bob Kerrey was the only commissioner who dared to disagree. "I do not believe it's an either/or choice of being angry at those who perpetrated this crime and feeling anger toward those with responsibility," he said. Kerrey later said, "I think the mayor should be held responsible for not having a plan for a crash." While Kerrey was the sole commissioner to raise questions about Giuliani's deflecting single-mindedness, staff director Philip Zelikow delivered a similar message in a radio interview in 2004. Asked whether the blame lay only with the terrorists, Zelikow replied, "Mayor Giuliani has expressed that very well. Of course, the people who are responsible for these deaths ultimately are the killers, but it's still our job to do what we can to mitigate their damage."[24]

Instead of mitigating the damage, the city's response added to it in ways far beyond the absence of working radios or preplans. Not only had the city done nothing since 1993 to prepare for an event as predictable as a mega high-rise fire, it acted on "instinct," as Sam Caspersen described it, once the inevitable happened. No coherent firefighting strategy emerged in the course of the nearly two-hour response. "You have to have a plan for major incidents and they didn't have one," Caspersen said. The deadliest consequence of this strategic void was that 500 to 600 firefighters climbed the stairs, even though the top chiefs had decided at the outset that all the department could do was a search and rescue operation. "The Fire Department errs on the side of putting too many people in harm's way than too few,"

Caspersen observed. "They were going to flood the tower. They had no idea of incident management. The thumb didn't know what the little finger was doing. They had no idea how many firefighters were there."

Caspersen doesn't think it was a matter of chaos caused by crisis; he ascribes it to "no standard-operating-procedure protocol beforehand." Cate Taylor agreed with her colleague: "Why were they sending all these engine companies up?" Taylor asked, pointing out that engine companies are supposed to put out fires. The World Trade Center operation had already been determined by Pete Ganci to be a rescue mission, the job of ladder companies. "We asked and never got an answer. It was not due to chaos. The chiefs were sending them up. There was a certain amount of control, a kind of fraternity thing, and everybody wanted to do something. There was no reason for all those engine companies. We got a lot of chaos-of-the-day type answers."

One of the reasons engine companies were ascending, however, was that chiefs were giving firefighters a mixed message, even though the top command had decided the fire could not be extinguished. "Some firefighters believed that this was primarily a fire-fighting mission," said Caspersen, "and others believed it was a rescue mission. A uniform, clear plan was not conveyed to each engine company and ladder company that ascended the stairs."

The National Institute of Standards and Technology described "a trinity of operational strategies"—with the top chiefs trying to evacuate those below the fires, the building chiefs hoping to cut a path through the fires to rescue occupants above the fires, and the company-level captains actually believing they could put out the fires. "These three operations strategies highlight differences that may be attributed to years of experience, level of training, and institutional focus of the various command levels," NIST said. "As the senior command level operational strategies were communicated to the lower levels, the concepts appeared to take hold at a slower pace at the next level down."

The top chiefs were "correct," the NIST report found, that the fires "were too large and too high" to be fought, especially with a ruptured water supply: "In some cases firefighters were persuaded by higher ranking officers to switch from the idea of firefighting to evacuation and rescue operations." But "some firefighters at the company level were disturbed by the operations orders that signaled a change toward assisting with the evacuation," and "wanted to go up and put the fires out." So, even though these "company officers were given information defining their mission" as evacuation, several continued to see "the event as a very large and dangerous yet still conventional high-rise fire" that they would find a way to extinguish.

Apparently, command within the department was so uncertain that only chiefs with talk-show host persuasiveness could control the situation. It wasn't just the 160 off-duty firefighters who rushed to the scene, propelled by their own fraternal loyalties. The chiefs and officers in charge of buildings and companies freelanced as well, putting firefighters on stairs with objectives contrary to the top command.

Frank Gribbon, the Fire Department spokesman on 9/11, offered the department's grisly calculus: "The commander at the scene would calculate the risks to the firefighters against the number of lives that could be saved." Of course, Ganci and lobby chiefs, as Caspersen points out, had no way to know precisely "how many civilians were trapped but in a position where they could be rescued." Caspersen conceded he couldn't estimate, "even with 20/20 hindsight, what would have been an appropriate number of firefighters for each tower." But Ganci and Feehan both said that they couldn't put the fire out or save anyone above it. The lobby chiefs "also knew from building personnel that some civilians were trapped in elevators and on specific floors," the commission report observed. With that information, wouldn't the narrow goal of rescuing the impaired or trapped require far fewer than 600 firefighters? Isn't it a job more for a platoon than for an army? And didn't this excessive number of ascending firefighters add to the radio mess, dramatically boosting transmissions that "stepped on each other" and consequently could not be heard? That's why a single, disciplined, definition of mission was so crucial to the rational dispatching of manpower. Instead, chiefs in the lobby were saying simply, "Do the best you can," to captains who checked in for instructions.

The doubts about the management of firefighters that day echo those of Tom Von Essen, who said he had been "convinced that we often sent too many guys into fires" since his "early days on the job." Von Essen told the *Times* in 2002: "I've been a firefighter since 1970, and have often stood on floors where we needed 10 people and had 30. There's a lack of control that's dangerous on an everyday basis to firefighters." Von Essen's admonition in his own book—that "the time may come when we have to hold back or limit our response"—was triggered by his desire "to protect the men who will respond to the next call." It was also a result of his belief that "too often, we sent men into fires without giving them all the training and discipline we could."

Traumatized by the mass grave of firefighters at Ground Zero, Von Essen said that he couldn't "stop asking questions," even whether "more men, even one man, might have been saved" had the department functioned better. "I don't think we can answer any of those questions conclusively," he decided. But he ruefully referred to the "temptation to postpone hard thinking and planning for such threats" and warned that "the next time we need to be ready beforehand."

Of course, the tragedy wasn't just the mistake of sending the men in; it was the botched effort to get them out, especially after the South Tower fell. Somewhere between 125 and 200 firefighters died inside the North Tower. In his testimony, Giuliani portrayed them as choosing to ignore the imminent peril in order to help civilians who otherwise would have been left behind on their own. "I know one firefighter whose family has explained this to me. He was in the North Tower. He was evacuating people. He was given an evacuation order. He told his men to go. He was with a person in a wheelchair, and an overweight person, so he stayed with them. So how did he interpret that evacuation order? He interpreted that order, 'I'll get all my men out but I'm going to stay here and help these people out.' And the fact that so many of them interpreted it that way kept [it] a much calmer situation and a much better evacuation."

There may well have been special circumstances—like Captain Burke's decision to stay with Beyea and Zelmanowitz—when a responder made that heroic calculation. But Giuliani is using an exception to conceal a failing. The National Institute of Standards and Technology found that half of the firefighters who died in the North Tower never got the order to evacuate. And of those who did hear it, many may not have understood its import. Of the 58 firefighters who escaped and did oral histories, 54 said they did not know that the South Tower had fallen, including many who heard a radio warning omitting that information. Sam Caspersen says that "the most important" Fire Department radio failing wasn't "technical" but was the "content of the message" on their radios. He names police captain Henry Winkler as one of the "top five heroes of the day" because, as soon as the South Tower collapsed, his instant evacuation order informed everyone on Police Department radios that it had toppled. Most Fire Department evacuation orders never said that; they just gave the order to evacuate without conveying the astonishing urgency of it. Though department evacuation protocols require a constant repetition of "mayday, mayday, mayday," that did not happen between collapses, either.

Instead of staying behind to rescue civilians—almost all of whom had already fled the North Tower by the time the South Tower collapsed—many firefighters were simply exhausted, collecting in large groups on the 19th floor and elsewhere. Even though the top chiefs had determined that the fires could not be extinguished, most firefighters went up in full firefighting gear, carrying over a hundred pounds, making themselves slower than many of the overweight and handicapped people they were sent to aid. Commission staffer Delgrosso was exasperated by the insistence of engine company firefighters that they "lug around fifty feet of hose" even though the buildings had their own hose. "There were also firefighter lockers every five floors with extra masks, air packs, oxygen, poles," said Delgrosso. "I can't tell you how many chiefs said: 'we wouldn't use their equipment.' The Fire Department didn't even know where the equipment lockers were. Other than the local engine company, no one else knew."

"A lot of companies had taken rests at about every fourth or fifth floor," one firefighter told NIST, "because we kept passing one another. It was very hot in the stairway, especially with all our gear and hood on, with no ventilation. I ended up going up to the 19th floor."

A Port Authority worker said, "I saw those firemen on the 27th floor. They were totally exhausted. They basically crawled onto the floor. That's why I ran and got them water." NIST found several cases of firefighters suffering from chest pains and heart attacks. During rests, the firefighters took off their turnout coats, which act as a thermal insulator and contributed to the physical drain.

Too many firefighters carrying too much equipment: it was a formula for fatalities. William Walsh, from Ladder Company 1, heard the evacuation order and walked down, pausing at the 19th floor to warn firefighters "hanging out in the stairwell" and on the floor. "I told them 'Didn't you hear the mayday? Get out.' They were saying 'Yeah, we'll be right with you.' They just didn't give it a second thought. They just continued with their rest." Of course, like Walsh, the resting firefighters had no idea the South Tower was gone.[25]

Deputy Chief Charles Blaich raised questions about the department's evacuation at a forum in January 2002, stirring a brief media furor. Praising the firefighters who went into the towers, he said, "There has to be a level somewhere saying, 'That's wonderful, now get out of here.' There has to be someone to make that hard call." Blaich said the department "lost track" of the firefighters who went up, affirming what the commission would find—namely, that "the FDNY as an institution proved incapable of coordinating the number of units dispatched." The only chief to publicly cast doubt on the management of the Fire Department evacuation, Blaich later claimed he was penalized for his candor.

In addition to the losses suffered because too many went in and too few got the word to get out, the National Institute of Standards and Technology report estimates that 160 responders died outside the towers, a particularly inexcusable loss. Firefighters gathered, for example, in droves at the undamaged Marriott Hotel, known as 3 WTC, adjacent to the North Tower and right in front of the South Tower. Even after both towers were engulfed in flame, at least 60 firefighters stayed in the 22-story hotel. Nearly an hour after dozens of buildings much farther away were emptied, responders, hotel staff, and civilians still milled around the Marriott, oblivious to the danger above. With its own separate and working power system, the Marriott allowed teams of firefighters to ride repeatedly to the top, theoretically making sure the 940 guests had left.

There were actually more firefighters in the Marriott than there were in the South Tower when it collapsed, though few had been sent to the hotel by command. In fact, Deputy Chief Tom Galvin, who was dispatched to the South Tower to take over as incident commander after the second plane hit, decided to get there through the hotel, and wound up getting "bogged down in there," as he put it in his oral history. He sent several units up to the floors to do another round of searches. Since the hotel was the "safest way" into the towers from West Street, according to Galvin, "everybody seemed to be coming in through it and there were a lot of companies in there."

The commission reported that 14 units were in the Marriott, while the South Tower, unlike the North Tower, was so short of companies that another alarm went out at 9:37 for more firefighters. When the collapsing South Tower cut the hotel in half at 9:59, it was an hour and 11 minutes after the Marriott began its own evacuation. There was no reason for 20 firefighters to still be on the upper floors, or for 16 of them to die. There was no reason for 50 civilians to still be in the restaurant. The hotel proved a doubly costly diversion when a second group of firefighters entered it to rescue those trapped under South Tower debris, and the North Tower collapsed on them.

Like a chain reaction pileup, the companies that were dispatched by the 9:37 alarm, which was prompted by the Marriott glut and the South Tower shortage, went to command posts and staging areas that were too close to the towers. They got there by 9:55 and suffered losses in both collapses, arriving too late to be of any use to anyone. While the location of these posts has been excused because the chiefs had no reason to anticipate unprecedented total collapses, the National Institute of Standards and Technology found that the same chiefs regarded "localized collapses in and above the fire zones" as "likely." Since a few falling floors would kill, too, "localized collapses" were clearly sufficient cause to set up posts farther away. Even Von Essen, who went to West Street, said later, "Certainly our command post was not far enough away from the scene as it should have been."

NIST found that "almost all emergency responder departments"—including police, fire, and Emergency Medical Services—"established their command posts within the potential collapse zone of the buildings" and that these decisions were consistent with their department procedures. It recommended that "command posts should be established outside the potential collapse footprint of any building which shows evidence of large multi-floor fires or has serious structural damage."

While the 9/11 Commission put no hard number on the out-of-tower responder deaths, it said many died in the Marriott, on the neighboring streets where the posts were, and on the concourse. Strangely, fleeing civilians also reported seeing many firefighters in the lobby of the North Tower shortly before its collapse, even though the South Tower had already fallen and the North Tower command post had long since been abandoned.

Instinct reigned in the minutes after the South Tower fell. Forty-three more units, with hundreds more firefighters, were dispatched to the scene between the South Tower collapse at 9:59 and the North Tower collapse at 10:29. Twenty-nine units arrived. Within seconds of being informed that "the entire building has collapsed," Manhattan dispatch was telling six engine companies to take the West Side Highway down to West and Vesey Streets, right next to the remaining tower. When the dust cloud in the Brooklyn Battery Tunnel lifted at 10:14, "all units" were sent to the South Tower site. At 10:26, two minutes before the North Tower collapse, fire units on the scene asked for "all the help we can get," and dispatch responded, "Ten-four. We have multiple units on the way in now." Port Authority Police Chief Joe Morris later said, "In a way, thank God the buildings went down when they did because you had 500 more people ready to go into the buildings."[26]

The courage contagion ran to the top of the department as well. Of the department's 32-member executive staff, 26 went to the incident, even though, reported McKinsey, "many had no role or responsibility" and might have been "better able to support" the effort "from a protected location with appropriate communication infrastructure."

CAPTAIN AL FUENTES, who went with Bill Feehan, Pete Ganci, and Ray Downey as they walked through the swirling dust of the South Tower's collapse, said later that they had to make their way "over multiples of ten, fifteen and twenty foot mounds of debris on West Street."[27] This group, too, was heading for the Marriott. As they walked toward it, a chief fleeing the hotel told Downey that there were civilians and firefighters trapped there. Ganci radioed back from the corner of West and Liberty, right beside the Marriott, asking for "two of my best trucks." Downey headed inside to try to help people get out, leaving Fuentes on the street to signal people when no debris was falling and it was safe to leave. The city's top technical chief, who'd gone out to Oklahoma City to help orchestrate the response to that bombing collapse, Downey disappeared into a hotel that had already become a tomb. Fuentes couldn't see him anymore and moved closer to the deathtrap himself just as the North Tower collapsed. He miraculously survived, but the brain trust of Downey, Feehan, and Ganci perished.

Crushed on West Street outside the Marriott while lifting rock in his bare hands to reach an obese woman, Feehan symbolized the department's historic reliance on action over direction, valor over cautious command. He died a hero, as did every firefighter who raced into the towers that day and never returned. All of them would have chosen survival, had they been given the information, equipment, plan, and instructions that could have saved their lives.

Yet, pressed at his private appearance before the 9/11 Commission staff about the command and control breakdown on September 11, Giuliani blithely stated, "It was not a major problem."

Chaos and a mere 102 minutes to act do explain some of these blunders. But so do truculence and eight years of inaction. Defend-in-place and get-to-the-fire were senseless bromides substituting for judgment that day, as well as in the years that led up to it. Firefighters and civilians died because others failed—both to plan and to execute.

Giuliani concedes, as he did before the commission in 2004, that the response was "not flawless, not without mistakes." But these are caveats, not confessions. He's never so much as hinted at anything he or his administration might have done that multiplied the casualties. It's as if the moment we met the enemy was the first time we could have considered any of the obvious: the trade center as a target; a high-rise fire plan; emergency communications systems that worked for firefighters and fire victims; strategies that took into account roofs, stairways, elevators, and the impaired.

The mayor's destructive speculation that the North Tower firefighters deliberately sacrificed themselves was the kind of eulogy that leads only to more eulogies. "Rather than giving us a story of uniformed men fleeing, while civilians were left behind, which would have been devastating to the morale of this country," he said, "they gave us an example of very, very brave men and women in uniform who stand their ground to protect civilians." The best rebuttal comes from Giuliani's own partner, Von Essen: "We also need to expand our conception of what a firefighter's duty is, beyond the image of the lone, heroic man facing the flame with only a hose and his own courage. It's a horrible reality that this new world will demand both great courage and the discipline to temper that zeal if necessary. Finding the right balance will be the never-ending duty of chiefs, captains, and every leader in the firehouse, even at the risk, sometimes, of upsetting their own people."

PART TWO
THE UNLEARNED LESSONS OF 1993

CHAPTER 3
1993—FORGOTTEN TERROR

NO ONE KNEW at the time how benign a harbinger it was, but the 1,500-pound bomb that shook the towers on February 26, 1993, was rated by the FBI as "the largest by weight and by damage that we've seen since the inception of forensic explosive identification" in 1925.[1]

It arrived in the 2,000-car parking garage beneath the World Trade Center in a yellow van, mistaken for one of the nearly identical Port Authority vans that frequented the world's largest, seven-building, complex. Inside the 10-foot van's 295 cubic yards of cargo space were four cardboard boxes of urea nitrate, a murky mix of fertilizer and acid bound together by wastepaper. Each box had its own 20-foot fuse, encased in rubber to slow the burn. Boosting each bomb was a nitroglycerine trigger, manufactured from 60 gallons of sulfuric acid, and a red tank of compressed hydrogen, a blast enhancer that had already become an international signature of Muslim extremists. A 99-cent cigarette lighter launched the new age of American terror a scant 12 minutes before the explosion—just enough time for Ramzi Yousef, Mohammed Salameh, and Eyad Ismoil to escape in a backup rented Corsica.

They had chosen their parking spot carefully, near a concrete wall almost midway between the two 110-story towers. When it went off, the bomb fractured the strong shoulders on which the vast complex sat. It carved a huge, jagged crater in the B-2 parking level, also destroying large swaths of the B-1 level above it. The crater was 150 feet by 130 feet and reached downward to B-5, nearly the bottom of a basement so vast it doubled all the square feet of the city's precursor giant, the Empire State Building. The smoke from the underground fires, fueled by the gas of at least 30 burning cars, jumped up the chimney-like elevator shafts to the top, overwhelming towers with tens of thousands of choking inhabitants. The force of 150,000 pounds per square inch threw a 3,000-pound steel beam 30 feet in the air like a jumbo javelin, sending it inside a North Tower lunchroom where four workers, one of them seven months pregnant, were instantly crushed.

Two million gallons of water gushed from severed pipes and air-conditioning chillers into the subgrade levels. Fifteen thousand square feet of steel and concrete were obliterated, leaving 2,500 tons of smoldering debris. A 200-foot section of underground ceiling collapsed on a commuter train platform below. Two hundred thousand linear feet of plaster disintegrated. Seven-inch masonry walls ruptured. An 11-inch-thick section of concrete floor, 80 feet long and 50 feet wide, fell two stories, coming to rest on a precarious perch below. A man at a stoplight 300 feet away from the towers saw his rear windshield shatter. A woman buying airline tickets three stories above the blast was blown 30 feet through the air. Six victims died and 1,042 were injured, 15 seriously. A federal prosecutor would later describe it as "the largest patient-producing incident in U.S. history apart from certain battles in the Civil War." It was also, no doubt, history's biggest building evacuation, with an estimated 50,000 visitors and tenants rushing toward the exits.

To the mastermind of the attack, however, it was a disappointment. After two months of mixing his deadly paste in Jersey City, a chagrined Yousef, whose dozen aliases make him an unidentifiable mirage to this day, studied his still-standing target at dusk from a waterfront vantage post across the New York Harbor. That night he would be on his way to Pakistan, mortified by the paltry death toll. When he was caught two years later, he boasted that his plan was to topple the North Tower onto the South Tower, sparking a firestorm of colossal death. His self-described goal was to kill 250,000 people. That's why the van was parked against a wall that separated the garage from a line of nine steel support columns, a quarter of the columns that held up the North Tower. Though the columns were severely shaken, the towers stiffened and stood.

But Yousef came very close to realizing his goal of mass murder on an epic scale. "The Vista Hotel, located directly above the blast area, was almost toppled," J. Gilmore Childers, a prosecutor who convicted most of the bombers, told a Senate hearing years later. The 829-room, 22-story hotel—located at 3 World Trade Center—teetered as if on stilts. Its basement columns, which had rested every 12 feet on solid concrete slab floors, were left without reinforcement for nearly five gutted stories. Without these floors, these new "long, spindly" columns were "much weaker than the short, stubby ones that they had been at the start of the day," experts concluded. Even Leslie Robertson, the structural engineer who'd designed the towers and toured them within hours of the blast, said the hotel columns "were subject to collapse at any time."

The Vista was still trembling as late as 9 P.M., when Police Commissioner Ray Kelly stood on a table in the dining room to talk to the 300 or so top officials and rescuers who'd gathered there for a briefing. The Port Authority's chief engineer whispered to Kelly's deputy that he was afraid the floor might collapse, since the crowd was right above damaged columns. But he was reluctant to interrupt Kelly.

"Tell him, tell him," implored the deputy.

"This is extremely serious. It is in danger of collapse," the PA official warned the group, and everyone relocated. Steel braces were quickly lowered through holes cut in the ballroom floor and fastened to shaky columns. Eventually, half an acre of cracked concrete floor slab was replaced.

But as Leslie Robertson roamed the cavern beneath the tower, he was even more concerned about the threat to the 10-acre foundation of the complex. Perhaps the WTC's greatest engineering feat was its buried slurry wall, a 70-foot-tall and three-foot-thick dike that literally held back the Hudson, whose tidal waters rushed up near the base of the complex. If that "bathtub" of reinforced concrete—3,100 feet around—was disturbed by the blast, the river would surge through the commuter and subway tunnels underneath the basement, wreaking havoc. One crucial concrete slab had already slipped three feet, requiring immediate bracing.

A third threat would go unmentioned for more than a year. When Salameh and his codefendants were sentenced in May 1994, U.S. District Court Judge Kevin Duffy declared, "You had sodium cyanide around and I'm sure it was in the bomb. Thank God the sodium cyanide burned instead of vaporizing. If the sodium cyanide had vaporized, it is clear that what would have happened is that cyanide gas would have been sucked into the North Tower and everyone in the North Tower would've been killed."

Senator Jon Kyl, chair of the Subcommittee on Technology, Terrorism and Government Information, put it this way: "Shortly after the bombing, FBI agents discovered in the storage locker of one of Ramzi Yousef's co-conspirators all the ingredients necessary for making poison gas hydrogen cyanide. This fact is well known. Less well known is that Yousef told FBI personnel who were accompanying him back to the United States after his capture that he had indeed originally hoped to explode a chemical weapon in the World Trade Center; the only thing that apparently stopped him was lack of funds to finance the operation." All he had to spend, according to prosecutors, was $8,000 to $12,000.

But it wasn't just a few thousand missing dollars that thwarted Yousef's deadly plan. He got off to a late start that morning. The plan was to detonate the bomb a couple of hours earlier, when the buildings were fully occupied. But the steel-nerved Yousef overslept. Instead, the 12:18 blast killed only the handful of basement workers who hadn't gone out for lunch. The FBI declared it a miracle that only six people died, and that was partly due to the fast work of the Fire Department.

The first FDNY units—Engine and Ladder companies 10—arrived in a single minute, merely crossing the street from their firehouse. In the end, 750 firefighters, on and off duty, wound up at the WTC. Eighty-eight firefighters, 35 police officers, and one emergency service worker sustained injuries. One firefighter, Kevin Shea, fell 45 feet into the dark crater, was rescued by another firefighter descending by rope, and survived major surgery. The basement fire was brought under control a mere hour and a half after the explosion. "The decision to attack the basement fires in the initial stages of the incident," wrote Anthony Fusco, chief of the department, "was the most important decision in that hundreds—maybe thousands—of lives were saved due to the timely extinguishing of the fires." No one died of smoke inhalation, concluded Fusco, precisely because the fire was rapidly suppressed.

Five hundred people were trapped in elevators, including 72 grade school students and teachers who spent five hours in one smoky car. Since the bomb shut down the Operations Control Center, which was located in the basement, the public address systems became instantly inoperable, leaving tens of thousands to make up their own minds about whether to stay or go. With regular and emergency power out, they fled down black stairwells that, especially right after the blast, were drenched in smoke. It wasn't until 5 P.M. that the first firefighter made it to the top floor, having worked his way up crammed stairwells barely wide enough to squeeze past the descending crowd.

"You could sense the fear in this city," Police Commissioner Kelly said.

NEW YORKERS ARE divided to this day about how good or bad a mayor David Dinkins was. What's inarguable is that he had terrible luck—a recession, a race riot, and a surge in the murder rate beset him almost from his earliest days at City Hall. When the World Trade Center was bombed, Dinkins, true to form, was far away in Japan on an economic development mission. His top deputy, Norman Steisel, rushed from City Hall to the WTC, where he operated out of a police-run mobile Fieldcom unit, taking a backseat to Chief Fusco, whose department was in charge, and Ray Kelly. Steisel woke Dinkins at 3:30 A.M. Japanese time, and the mayor immediately began his long trek back to New York, first by train from a city outside Tokyo and then aboard the earliest plane he could get a seat on. Meanwhile, Steisel, bald, grim, and pudgy, wound up doing national television interviews all day and night, making the kind of visceral connection with a panicked city that only crisis can confer. Used to the obscurity of a behind-the-scenes handler, he was stunned a few days later when he and his wife went to one of the city's finer restaurants and got a spontaneous standing ovation.

Dinkins, who was up for reelection that fall, had missed a rare opportunity to make an I'm-in-control impression on a temporarily attentive populace. The 65-year-old mayor did not actually get to the bomb site until Saturday afternoon, 24 hours after the blast. Though it was never reported, Rudy Giuliani, who'd narrowly lost to Dinkins in 1989 and was running again, rushed to the site in his campaign car, but was barred by cops from entering. In public, Giuliani never made Dinkins's awkward absence an issue. But in private, he and his campaign aides derided the mayor, poking fun at the fact that he wore a construction helmet when he toured the site, even though it was long after the danger had subsided. They saw Dinkins's appearance as "lame" and "ineffectual."[2]

The mayor was also tormented by his Mario Cuomo problem. Dinkins was always upstaged by the charismatic governor, the reverse of what would happen in the aftermath of the second WTC attack on 9/11. Shortly before Dinkins got back on Saturday, Steisel hosted a nationally covered morning press conference. Before it, Kelly and FBI chief Jim Fox urged Cuomo, Steisel, and the rest of the officials not to tell reporters that the incident was a bombing. The earliest news reports had suggested it might have been a transformer accident, and Kelly and Fox wanted another day of muddled messages, apparently to decoy the perpetrators. But at the conference Cuomo couldn't contain himself: "It looks like a bomb. It smells like a bomb. It's probably a bomb." Then, as he left the mike and passed Fox, he smiled: "Well, I didn't say it was a bomb." With Dinkins just minutes from arriving, Cuomo had delivered the sound bite that would dominate the news cycle. "All of America has been violated by this," Cuomo added. "What it does to the psyche is frightening. We have lived behind this invisible but impenetrable shield for all these years."

If the public was eager to believe the impenetrable shield still existed, the swift arrests of the bombers helped the process of denial along. Salameh, who was first, became an instant national joke when, six days after the bombing, he returned for his $400 deposit on the Ford Econoline van he'd rented from a New Jersey franchise. Raids at Brooklyn and Jersey addresses listed for Salameh led to the busts of three co-conspirators and the identification of others, including the already vanished Yousef.

The FBI soon began ridiculing this clumsy crew, with officials calling them "unsophisticated" and *Time* magazine dismissing them as "rank amateurs." But the truth, which only emerged over time, was that this powerful if crude bombing was as much an indictment of FBI preparedness as the subsequent arrests were a measure of its investigative prowess. Eight months before the attack, the agency had fired an Egyptian confidential informant so plugged into this deadly circle that he'd warned the FBI about a bombing plot and urged them to tail two of the bombers, possibly even specifying the WTC as a target.

After the bombing, the FBI rushed to put the informant, a 43-year-old former military officer named Emad Salem, back on the payroll. Though he wanted only $500 a week before the attack, the FBI was now promising him $1.5 million to risk his life back inside the cell. "If Emad had stayed with us, yeah, he would have prevented" the bombing, said Lou Napoli, a police detective assigned to the FBI's Joint Terrorism Task Force in New York. In addition to Salem's warnings, a Palestinian terrorist group predicted in late January that "a bomb would be planted in a New York City skyscraper" that month, a threat relayed to the police. Pete Caram, the antiterrorism chief at the Port Authority, got a similar warning from the Joint Terrorism Task Force.[3] Security was soon beefed up at the WTC, but when no attack occurred, the alert was rescinded. Despite this fumbling, no one criticized the police, the Port Authority, or the FBI—not even mayoral candidate Giuliani, who never so much as mentioned the new terrorism, or cited the bombing as an example of it, throughout the 1993 mayoral race.

Charles Schumer, the Brooklyn congressman who would later

become a senator, certainly wasn't ignoring the security issues raised by the bombing. Calling it "a shot across the bow importuning us to act," Schumer's Subcommittee on Crime and Criminal Justice held a Washington hearing that posed a critical question: "Are there reasonable precautions we can take to improve the security of our public buildings?"

The Schumer hearing—like much of the "new era," postbombing press—highlighted the continuing nature of the terrorist threat in New York, and especially at the World Trade Center. Police Commissioner Ray Kelly testified: "Obviously, we're concerned about the possibility of the bombing being the first in a series." David Dinkins—in one of the mayor's rare public discussions of the subject—told the committee that "our city and our world have drastically changed" and acknowledged that "we in New York and we in America must rethink our national attitude toward terror and security."

Brian Jenkins, the senior managing director of Kroll Associates, the international security firm retained by the Port Authority after the bombing, observed, "With this bombing, a taboo has been broken. Others may be inspired to follow the example." Neil Livingstone, director of the Institute on Terrorism, told the committee that the World Trade Center remained "one of the three best known structures in New York City," along with the United Nations and Statue of Liberty. Calling terrorism "a form of communication" utterly dependent on publicity, he prophetically noted that "by hitting" the towers, "the terrorists were virtually guaranteed that there would be a vast amount of electronic and print coverage."

A COUPLE OF weeks after Schumer's autopsy of the attack, State Senator Roy Goodman, who doubled as chair of the Manhattan Republican Party, started three days of hearings on March 22, designed to examine "the security and safety aspects" of the WTC bombing. Goodman's committee questioned 26 witnesses in a no-holds-barred probe of the city and of the Port Authority, both Democratic bastions. He opened the hearing by branding the bombing a "tragic wake-up call" and a "dire warning of future disasters with far greater loss of life if we fail to prepare" for terrorism "here at home." He laid bare a shocking series of Port Authority consultant and internal security reports that predicted exactly the kind of garage bomb the terrorists pulled off. He berated one Port Authority executive director after another about why the security recommendations weren't followed, particularly when it came to the wide-open parking garage. He likened one witness, Edward O'Sullivan, to the seer Nostradamus and told him that he was "thunderstruck at the degree to which your crystal ball was functioning" in 1985, when his 120-page study pinpointed the vulnerabilities exploited by the bombers.

Another persistent in-house advocate for closing or tightly regulating the garage was Pete Caram, the only Port Authority employee with a top security clearance and Joint Terrorism Task Force assignment. Caram sounded an alarm at Goodman's hearing, saying he feared "further disaster somewhere down the line." He urged the authority to "harden our target," meaning the World Trade Center. "I can't see implementing counter-terrorism programs on a crisis basis only," he concluded. Caram's warning was echoed by the FBI's Fox, who somberly concluded, "We would be well advised to prepare for the worst and hope for best." Kelly said the city should be at "a heightened state of awareness and readiness for the foreseeable future."

Goodman quickly introduced a bill in Albany to bring Port Authority buildings, including the World Trade Center, under the jurisdiction of the New York City building code. The bi-state authority had long exploited its exemption from any local code, cherry-picking the standards it would meet, especially at the trade center. Port Authority officials prattled on for years about how they voluntarily met or exceeded city code, but everything from the fireproofing to the stairways to the wide-open floors sidestepped it. The Port Authority's exemption meant it didn't have to file permits with the city buildings department, so there was no public record of what part of the code it chose to adhere to and what part it chose to ignore.

The Port Authority was determined to head off Goodman's bill. Stan Brezenoff, the Port Authority's executive director, decided that the best way to defend the exemption was to look like he was surrendering to it. He announced on the eve of Goodman's hearing that he would execute memos of understanding with Dinkins's buildings and fire departments that committed the authority to code compliance and granted oversight powers to the city. The agreements wouldn't actually be signed until late 1993, and, while they did strengthen city powers, the authority still held the upper hand.

The Fire Department did get some limited right to review and approve fire safety system projects and to do inspections. But Chief Bill Feehan, the FDNY's first deputy commissioner, testified that "the problem" with memos of understanding is that all they empower the Fire Department to do is "act as their consultant." When the Fire Department and the Port Authority disagreed, said the then 34-year veteran of the department, "they are not required to accept our advice." Feehan submitted a statement from Fire Commissioner Carlos Rivera, a Dinkins appointee, who said it was his "firm belief" that legislation should be passed to ensure that Port Authority buildings comply with codes and giving his department enforcement power at the World Trade Center. Brezenoff said he understood the department's concern and insisted the memos would change all that.

When that exchange was over, Goodman asked if it was the "overwhelming sentiment" of the fire and buildings departments that the authority should be compelled by law to comply with the city code, and Feehan and Buildings Commissioner Rudolph Rinaldi said yes. A deputy mayor appeared at the hearing and endorsed the senator's bill, while Norman Steisel promised that city lawyers would draft their own version within 45 days. But nothing happened. When Goodman released a final report that August, contending that his bill would "at least insure that the City is never again so unprepared" for a terrorist event, and defining the WTC as "a singular potential terrorist target," the two mayoral combatants, Dinkins and Giuliani, were silent.

Giuliani's silence was the strangest. Goodman was, in many ways, his mentor, intimately involved in the 1993 campaign. Giuliani was a member of the senator's East Side Republican Club and had announced his candidacy for mayor there in 1989. As the GOP leader for Manhattan, Goodman raised money for Giuliani and supplied foot soldiers. Yet when Dinkins walked away from an action plan he'd once embraced, Goodman couldn't get his own candidate to take up the cause. Seth Kaye, an ex–Goodman aide who became a key campaign adviser to Giuliani, even brought up the Goodman WTC findings and bill within the campaign staff, according to other policy aides. But it all fell by the wayside. Eight years later, Goodman would say his hearings and report were "ignored," but he drew a blank on any interactions he might've had with Giuliani about it.

At the same time as the well-covered Goodman hearings, Giuliani was quietly participating in his own carefully crafted, private meetings with experts on the issues of municipal governance he knew so little about—from schools to the budget. On March 25, sandwiched in between the two days of Goodman testimony, Giuliani met for the first time with Bill Bratton, the Boston police commissioner who would ultimately replace Kelly. The interview occurred less than a month after the bombing, but the subject never came up. Neither did the subject of terrorism, in any form.[4]

Bratton had been the Transit Authority police chief in New York before moving on to Boston, pioneering strategies of zero tolerance for fare beaters and graffiti, demonstrating that a crusade against quality-of-life infractions in the subways could also cut the felony crime rate. Giuliani wanted to know if the same strategies would "work on the streets" of the city. Even though crime had declined throughout the last three years of the Dinkins administration, New York was still widely regarded, especially by its citizens, as a city out of control. Giuliani knew his candidacy depended on whether voters could be convinced that he would make them safer—not from Muslim extremists but from muggers and crazed homeless people.

It was hardly surprising that the candidate stressed these issues in his three-hour session with Bratton, or that he wasn't planning to run on terrorism. What the interview indicated, though, was that the new terrorist threat was nowhere on his public safety radar screen. Giuliani's campaign staff remained in phone contact with Bratton throughout 1993, peppering him with periodic questions about policing, but they never raised a question about terrorism or the bombing.

And when it came to terrorism, there continued to be a lot to talk about. On June 24, the FBI raided a warehouse in Queens where five Muslims in white overalls, who were linked to the WTC bombing, were busted with plans—and enough fuel and fertilizer—to blow up two Hudson River tunnels, the United Nations, and the tower that housed the FBI. The banner on the *Daily News* was: TARGET N.Y.: ON THE BRINK OF TERROR. The FBI's Jim Fox also announced the arrest of three other conspirators, adding that the bombers were caught with five metal drums filled with a "witches brew" of explosives. Kelly noted that "they were looking" to execute their "Day of Terror" plan "in the next week."

The million-dollar informant Salem had picked the small warehouse on 90th Avenue in the Jamaica section of Queens without telling his co-conspirators that it was wired for sound and video. In the following weeks, the number of conspirators indicted grew to 13, led by the infamous blind sheik, Omar Abdel-Rahman.

Newsweek's 2,500-word account of the bust was typical of the news coverage: "The possibilities were stupefying—too dreadful to be believed. Some day very soon, perhaps as early as this week, New York would be struck by a series of powerful explosions that would bring the city to its knees. First, bombs would go off in the Holland and Lincoln Tunnels, trapping thousands of commuters deep below the Hudson River in the ultimate claustrophobic horror. Two more explosions would follow: at the United Nations building and the Jacob Javits Federal Building. Swamped by multiple catastrophes, police, fire and ambulance crews would be stretched beyond their limits. New York would panic." Had the gang not been penetrated by an informant, *Newsweek* declared, its "plan almost certainly would have worked." The arrests and the WTC bombing "demonstrated all too well," the magazine concluded, that "New York and many other 'soft targets' across America are now undeniably vulnerable to terrorist attack."

The drumbeat of local and national coverage—much of it focused on the sheik—made terrorism the hottest story of the day, all of it with implications for the next city administration. But even when it was revealed that Giuliani's top Orthodox Jewish supporter, Democratic assemblyman Dov Hikind, had been under 24-hour police protection for more than a month because of death threats by the same terrorist cell, the campaign said nothing. Hikind, U.S. Senator Al D'Amato, and City Comptroller Liz Holtzman, embroiled in her own reelection race, publicly demanded the arrest of the sheik for days before Attorney General Janet Reno finally acceded to it. But Giuliani said nothing.

Even when the trial of the first five WTC bombers started in mid-September, capping off what would be the worst year of at-home terrorist attacks in history, Candidate Giuliani saw no need to discuss city preparations for future incidents. Though his law enforcement credentials and Dinkins's halfhearted response represented a political opportunity, Rudy stayed on message, and his message did not include terrorism.

Richard Bryers, who was Giuliani's campaign spokesman, acknowledges that they "did absolutely nothing about the bombing" or the warehouse conspiracy, saying they were focused on the murder rate, budget, and racial tensions. "We did three weeks on whether it was appropriate to use the word 'pogrom' to describe" the anti-Semitic riot in Crown Heights. "We didn't make any commercials about the attack; it never came up internally. It was seen as a foreign policy issue. Campaigns tend to focus on the top concern of the moment. Terrorist concerns were not a top concern." The only reference Giuliani made to the bombing came in a major September 9, 1993, speech about quality-of-life issues, when he used it as an inspirational prop. "What I will do is ask you to work with me to restore our city," he said, "to reach down and show that same kind of spirit and grit that we showed during the World Trade Center explosion."

THE DISCONNECT BETWEEN Giuliani and counterterrorism—both as a candidate and ultimately as mayor—was all the more surprising considering his ostensible career connections to it. In a 15-year career as a G-man, Giuliani ran the entire criminal side of the Justice Department in Washington from 1981 to June 1983 and served as U.S. attorney in Manhattan from 1983 until days before he announced his first mayoral campaign in 1989. While terrorism was hardly a high-profile issue in those years, Giuliani has not hesitated, before and since 9/11, to suggest that his federal law enforcement background equipped him in an unusual way to both understand and combat terrorism. In a 1998 television appearance, he said, "I had to deal with terrorism when I was U.S. Attorney, and when I was the third ranking official in the Justice Department, so I have a pretty good sense of it." And in his post-9/11 book, *Leadership*, he reiterated how his "career as a prosecutor" had prepared him to put himself "inside the minds" of terrorists.

The highlight of that early resume was his claimed frontline status in the fight against Yasir Arafat and the Palestine Liberation Organization. As early as his 1989 mayoral race, well before any major attack on the city, Giuliani branded Dinkins "soft on terrorism" because Dinkins wouldn't oppose an Arafat visit to the United Nations. Giuliani declared then that as mayor, he would have Arafat "arrested as an accomplice to terrorism" if he entered the city. He also made much of the fact that he had "personally argued the case to throw the PLO out of this country," a reference to an instantly dismissed eviction proceeding filed against the PLO's mission in 1988. In a city where the subject of international relations often shows up in mayoral campaigns, Giuliani's stance was not regarded as anything other than the politically pragmatic antipathy toward enemies of Israel that is as commonplace in New York as handshakes and cheek kisses. (Although Giuliani never carried out the campaign promise to have Arafat arrested, he did create a diplomatic incident in 1995 when he had Arafat thrown out of a Lincoln Center celebration of the United Nations' 50th anniversary.)

Giuliani claimed his hatred of Arafat stemmed from the days when "I investigated Arafat for his alleged involvement in the murder of Leon Klinghoffer." He was referring to the 1985 murder of a New York businessman aboard the *Achille Lauro*, an Italian ship hijacked off the coast of Egypt by Palestinian extremists. The killing was so savage—the wheelchair-bound, 69-year-old Klinghoffer was shot in the head and dumped in the Mediterranean—that, as Klinghoffer's daughter would later say, it "put a face" on terrorism in the American mind. Giuliani has repeatedly cited his supposed work on the Klinghoffer case since the

'80s, even including it in his book, *Leadership*. The lionizing historian Fred Siegel, who was a policy adviser to Giuliani in his mayoral campaign, went so far as to assert in a TV interview "the little known fact" that Giuliani "would have been the one to prosecute" the hijackers "had they been brought back to the U.S." As mayor, Giuliani cited the Klinghoffer case as the reason for his "special contempt" for Arafat, adding that his probe in the '80s had uncovered "evidence of maybe 30 or 40 people he had murdered."

Yet there was never any criminal investigation by Giuliani or the Justice Department that directly implicated Arafat in the Klinghoffer killing. Giuliani's friend and mentor Arnold Burns, who was associate attorney general in Washington when the *Achille Lauro* case was reviewed there and who argued a civil case on behalf of the Klinghoffer family before the U.S. Supreme Court after he left government, says Giuliani was never involved with it. Burns, the finance chair of Giuliani's first campaign for mayor in 1989, added, "I know of nothing Rudy did in any shape or form on the Klinghoffer case." Jay Fischer, the Klinghoffer family attorney who spearheaded their litigation for 12 years and eventually won a monetary settlement with the PLO, says he "never had any contact" with Giuliani or anyone in his office about the case. "It would boggle the mind if anyone in 1985, 1986, 1987 or thereafter conducted an investigation of this case and didn't call me," said Fischer.

Victoria Toensing, the deputy at Justice who did investigate the Palestinian leader who organized the *Achille Lauro* hijacking, Abu Abbas, says that no one in Giuliani's office "was involved at all." Asked if there was any consideration given to indicting Arafat over the Klinghoffer case or other terrorist attacks on Americans, Toensing said, "Sure, people were always talking about indicting Arafat. I explained to those people that they had to show me exactly what law he had broken." (Abbas represented his own organization on the PLO executive council).

The only involvement Giuliani really had with Arafat was the eviction suit almost three years after the Klinghoffer affair, and he's hyped his role in that as well. It was Justice officials in Washington who decided to file the suit, spurred by legislation in 1987 that appeared to mandate it. Arnold Burns says he "assigned Rudy to shut down the PLO office in New York." Together with John Bolton, then head of Justice's Civil Rights Division, Giuliani brought the case in Manhattan, unrestrained by the U.N.'s 1974 decision to grant the PLO official observer status. The U.N. General Assembly voted 148 to 2 to oppose the Justice action, Secretary of State George Shultz called the eviction "dumb," the World Court ruled unanimously against it, and even incoming president George H.W. Bush "distanced himself" from it, according to news accounts. But Giuliani still attempted to have the utilities at the PLO's East Side townhouse turned off, ultimately losing the case and abandoning an appeal he announced he would file.

Beyond the flawed eviction proceeding, the only terrorist case Giuliani ever brought to indictment was announced with great fanfare at a 1986 press conference, and had nothing to do with Arafat. It involved a scheme to sell $2.5 billion in antitank missiles, fighter jets, bombs, and other hardware to Iran. Giuliani called it "mind-boggling" and the biggest arms smuggling case in American history. U.S. Customs Commissioner William Von Rabb called the 10 defendants "brokers of death who operated a terrorist flea market." The sale was to the Iranian government, but on Giuliani's motion, a judge ruled that it was the legal equivalent of a terrorist sale, since Iran was then tied to 87 terrorist incidents over the previous three years. The case never became part of the now much-vaunted Giuliani counterterrorist resume because he personally filed papers terminating it in his final month in office in January 1989. While the mysterious death of a key witness hurt the case, it was also crippled by errors in Giuliani's office, with 46 of the 55 charges stricken by the courts.

When he left the Justice Department to start his political career, he worked as a corporate lawyer—for longer than he had intended, since he lost his first race against David Dinkins in 1989. The law firms he wound up working at did business with multiple institutions that had ties to terrorist activities. The firm he joined in 1989—White & Case—became a campaign issue in part because of its representation of the Saudi-based Bank of Credit and Commerce International (BCCI), a veritable life force for drug cartels, despots like Saddam Hussein, and even the most notorious terrorist of the time, Abu Nidal. This was merely the crassest of the law firm connections that may have also partially explained Giuliani's reluctance to talk about terrorism in the '93 campaign. Giuliani personally represented the Milan-based Montedison, which was accused in Italian press reports of providing "chemicals and expertise" to a poison-gas plant in Libya. Already on the U.S. terrorist state list while Giuliani represented Montedison, Libya also began surfacing as the possible sponsor of the bombing of Pan Am 103 over Lockerbie, Scotland, an attack that killed 217 Americans.

His second law firm, Anderson, Kill, Olick & Oshinsky, represented 11 companies, mostly European, accused of selling components of chemical weapons or other arms to Saddam Hussein.[5] Giuliani represented one himself. He also had an opportunity to object to the rest, since a 1991 memo was circulated within the firm listing the clients of a prospective new partner, and most of them were on it. The firm was debating whether to accept the partner, and the Iraqi connections of some of the companies had been published by the *New York Times*, provoking American Jewish leaders to meet in protest with the German chancellor about the most egregious examples. Neither Giuliani nor any of his other ex-prosecutor friends at the firm, including Denny Young, said a word. At the time of the bombing in 1993, Giuliani was still running his fledgling mayoral campaign out of a conference room at a firm festering with strange Saddam associations. One of the bombers, Aboud Yasin, fled immediately to his native Iraq, where Hussein welcomed and harbored him, feeding neoconservative suspicions to this day that Hussein had sponsored the attack. As this awkward nexus of career and coincidental connections emerged, Giuliani remained, as he had at White & Case, tone deaf to the rumble.

ONCE GIULIANI WON in November of 1993, the immediate focus of his postelection transition was the selection of a police commissioner. Kelly and Bratton appeared in news accounts to be the prime candidates, but Kelly says now that he wasn't seriously considered. An ex-marine, Kelly was championed with Giuliani by another ex-marine,

Republican Staten Island borough president Guy Molinari, one of Giuliani's most influential early supporters. "Guy set up a meeting for me with Giuliani," recalls Kelly. "It was kind of mysterious. It was in a hotel room, super secret, and Giuliani mostly asked me questions about merging the transit police with the NYPD. We also talked about street crime, although not in depth. He did it to keep Guy happy. There was no discussion about terrorism or February 26. Then I met with a committee of Giuliani's and they never mentioned the bombing or anything about terrorism either."

The mayor-elect's public safety transition committee interviewed several other candidates as well. One was Henry DeGeneste, the former Port Authority superintendent of security who had moved on to become head of global security operations for Prudential. Another was Leslie Crocker Snyder, a Manhattan Supreme Court judge with an impressive record in the district attorney's office. Neither recalls even a single question about terrorism or the bombing. DeGeneste says he'd "written articles and spoken about counter-terrorism," but that he was only asked about drug trafficking and community policing. Kelly, who helped spearhead both the WTC and the warehouse bombing probes, remembers clearly that "no terrorism-related issues were raised." Snyder is "quite sure it never came up," as is Bratton. "These guys had no understanding of the threat of terrorism," says Bratton, "they were concerned with day-to-day crime and homicide in the streets."

Bratton was initially interviewed the day after Thanksgiving at the Scarsdale farmhouse of Adam Walinsky, the former chair of the State Commission of Investigation and a member of the transition committee in charge of police, fire, and corrections. On the same day, Walinsky and Howard Wilson, a former federal prosecutor close to Giuliani and chair of the committee, took Bratton to a secret session with Giuliani at Wilson's law office. Bratton then appeared before the eight-member transition committee and, finally, did an intimate session in December with Rudy and an inner circle of his five closest advisers that ran until 1 A.M. In all of those combined hours, there wasn't so much as a whisper about the terrorist threat that had so dramatically surfaced only months earlier.

"I don't believe that any member of the committee or anyone else involved spent five seconds, there wasn't five seconds of discussion, about terrorism," Walinsky recalls, adding that Giuliani never mentioned it as a factor in the selection process. "I don't believe one person in those meetings raised it." Walinsky and two other members of the committee—Herman Badillo, who ran for comptroller on Giuliani's ticket, and attorney Gary Cooper—agree that the WTC attack and the June breakup of the bombing plot never figured in the transition's public safety talks.

It did not even come up when the panel interviewed Tom Sheer, the former head of the FBI's New York office, who'd helped reorganize it to focus on international espionage. *Newsday* said the Sheer interview "was a courtesy to FBI director Louis Freeh," an old friend of Giuliani's who'd recommended Sheer. While Sheer did not have the urban policing background for the top job, the mayor didn't consider creating a counterterrorism deputy post for him, similar to the two created by Kelly when he became commissioner again under Mayor Bloomberg. Had Giuliani seen the terrorist events of 1993 as a rationale for such a position, he would have had a high-level conduit at the Police Department, connected to the FBI and federal prosecutors, focused on keeping the issue on City Hall's agenda.

A month after Bratton's selection, when Giuliani was sworn in on January 2, 1994, he became New York's first mayor in 100 years to take office without ever holding any other elective office. He was the stern and resolute embodiment of law enforcement; all he'd ever done was prosecute. His campaign revolved around a single promise: I will make you safe. He was, however, as disengaged from the new terrorist threat as the Harlem clubhouse incumbent he defeated. His city had been shocked and battered, not just by street thugs, but also by a shadowy and sinister new circle of mass murderers. Yet Giuliani came to City Hall as if this latest lethal breed did not exist.

The federal judge Giuliani asked to swear him in was Michael

Mukasey, a lifelong friend who'd tried a career-making congressional corruption case with him in the early '70s when both were young prosecutors. Only a few months before the inauguration, Sheik Rahman, the spiritual founding father of Islamic terrorism in America, and a dozen of his followers, had been arraigned before Mukasey, manacled, and charged with chilling conspiracies, the targets of which ran from the U.N. to the Hudson River tunnels. Mukasey would soon put Rahman and the rest of his Day of Terror gang in prison, some for life. If Giuliani needed a lesson in the gravity of the threat, all he had to do was read his close friend's statement to Rahman at the sentencing. "You were convicted of directing others to perform acts which, if accomplished, would have resulted in the murder of hundreds if not thousands of people," Judge Mukasey said in a courtroom that his friend Rudy made sure was heavily guarded. If the terrorists had been successful, he said, the bombings would have caused destruction on a scale "not seen since the Civil War."

That was the intimate law enforcement network from which Giuliani came. It was Giuliani, as U.S. attorney, who hired and befriended the men who made these cases, from Gil Childers to Rahman's prosecutor Andrew McCarthy, to Yousef's prosecutor Dietrich Snell, to the embassy bombing trial attorney Patrick Fitzgerald, to the chief of the counterterrorism unit David Kelley. Yet as mayor in the new age of terror, he never talked to them, never sought a briefing that might have allowed him to benefit, at least in broad strokes, from the unique understanding of this underworld that only a prosecutor who's tried one of these complex cases has.[6]

At the 9/11 Commission's final hearing in 2004, former U.S. senator Bob Kerrey suggested just how valuable such briefings might have been while questioning Fitzgerald, the assistant U.S. attorney who spearheaded the Yousef, Rahman, and embassy bombing investigations. If Fitzgerald "had gone and briefed" President Bush, delivering just "the public information you had at trial," Kerrey said, Bush would have learned far more than he got from the infamous CIA daily briefing on August 6, 2001. Claiming that he'd now seen all the classified material, Kerrey declared there was "more content" in the court files about "what bin Laden intended to do," calculating that 70 percent of what the government knew was public by 1997 and 100 percent well before 9/11. Fitzgerald agreed: "I think it's fair to say that there's a lot in the open sources that wasn't reported widely. I've always been confused by why people didn't pay attention to what becomes public." A memo on bin Laden, for example, was filed at the courthouse as early as 1995.

U.S. Attorney Mary Jo White, who ran Giuliani's former Southern District office throughout his mayoralty, agrees with Kerrey, contending that "by 1998, all of it was in the public record." Yet, White added, neither she nor any of the trial assistants were ever asked to brief him or his staff. In the summer of 1994, White recalled, she invited Rudy to address the whole office, and he came, sat at his old desk, and gabbed before delivering a speech and taking questions. He talked about "how to make the city safer" and cited the gang cases the office was doing. "I don't recall the subject of terrorism coming up," White said. Her chief assistant, Matt Fishbein, remembered that 10 or 12 people, from White's and Giuliani's top staff, talked for a while before the speech and no one mentioned the bombing or terrorism.

White had taken over the Southern District in mid-1993, right before the Day of Terror busts. When this meeting occurred, her office had already prosecuted the first group of 1993 bombers and was trying the warehouse plotters. Yet Giuliani did not attempt to set up a pipeline of information to an office that would become the national headquarters for the war on bin Laden. The only time he ever discussed the 1993 bombing with White was when he called to congratulate her on the first convictions. White remembered two other meetings she had at City Hall with Giuliani, and believed that one involved the mob-infested Fulton Fish Market and that the other focused on city assistance for a youth program her office was sponsoring in Washington Heights.

"We didn't have any substantial discussions about terrorism until after 9/11," White said, saying that the sit-downs she did have with him were not "heavy duty" and "did not include" any exchanges about al Qaeda, the attacks, or the threat. White was also sure that no one from her terrorism unit—the only U.S. attorney's office in America to have one—was ever asked by the police or City Hall to "sit down and focus" on what it knew. White assumed that all Giuliani wanted was periodic intelligence about specific threats or events, not a broader terrorist picture, and she believed he got that from the FBI. "He could have picked up the phone" if he needed a fuller context, said Fishbein. "We were more on top of this than anyone else." Counterterrorism chief David Kelley said his highest-level interaction was with a deputy police chief, and that he relied on the detectives in the Joint Terrorism Task Force, which he oversaw, as the "conduit available to city government." He acknowledged that the city never increased the number of Task Force detectives before 9/11,[7] and that he didn't know "how the information went up the ranks" and simply "assumed" Giuliani was briefed, at least about specific threats.

White did not find the lack of substantive exchange surprising, even though her office was pursuing a cabal that menaced the city and Giuliani was charged with protecting it. Childers, McCarthy, and Snell did not regard it as any failing on Rudy's part that his administration had not tapped into their understanding of the threat, or even that no one from the NYPD had monitored their trials or sought their transcripts. But to anyone outside these two cocoons, the disengagement is mystifying. Certainly, Giuliani gave voters reason to believe they would benefit in unusual ways from the experience he frequently cited, and his fraternal law enforcement relationships were presumed to be part of that advantage.

It's not like there was nothing the new mayor of New York could do about the rising threat. The postbombing reports by Chief Fusco and other Fire Department brass were a mandate for reforming the city's response to a catastrophic attack. The Port Authority's admitted system failures were an invitation for aggressive city oversight and inspection. The litany of ignored warnings was a cry to better connect intelligence and action. Senator Goodman's bill was a recipe for jurisdictional empowerment. But with most of the bombers in jail and the towers reopened, the casualties and near-casualties of 1993 were already fading from view, especially the view from City Hall.

Just weeks before the first anniversary of the WTC bombing, Giuliani opened his City Hall inaugural with the fierce vow that "the era of fear has had a long enough reign." Though the Metropolitan Correctional Center, also just a couple hundred yards from where he stood, was already jammed with new jihadists, some on trial as he spoke, he did not include them in his definition of the fearsome. His key staff and new police commissioner were already exchanging memos detailing plans for the fight against fear, but these rigorous to-do lists did not mention terror. A summary of the initiatives discussed at a meeting with Bratton just eight days after the inaugural—prepared for Giuliani special adviser Richard Schwartz—spelled out the NYPD response to nine City Hall proposals without a reference to any counterterror strategy. A half dozen similar initiative memos between City Hall and the Police Department in 1994 document the Giuliani blank slate on emergency terrorist response, without so much as a sentence about the bombing or the continuing threat.[8]

For only the second time, his speech that inaugural day did mention the bombing. But as he had in September, Giuliani reduced it to a self-help metaphor, converting the evacuation into a tribute to personal responsibility. "Fifty thousand New Yorkers took charge of themselves and each other," he declared on this chilly day of stiff resolve, anticipating the welfare and other reforms he would soon introduce. "The New York spirit that brought us through the World Trade Center crisis was a demonstration of the courage and ingenuity that we must now apply to restoring public safety, saving our schools, reducing our deficit and improving the quality of our lives."

In a memorable speech that challenged his city to change in so many other ways, Rudy Giuliani, 49, embarking on the greatest public mission of his life, described the trade center crisis as if it were over—proof, as he put it, that "nothing is beyond our grasp."

CHAPTER 4

A TRADE CENTER BESIEGED

FOR THE TERRORISTS, the World Trade Center was always an obsession.

The day after the 1993 attack, 25-year-old Nidal Ayyad, a Rutgers graduate, American citizen, and $35,000-a-year chemist at a major New Jersey company, began buying chemicals for another bomb. Ayyad was the first of the small, fanatical circle of WTC bombers to move toward serial killer status. The mastermind of the attack, Ramzi Yousef, was on a plane for Pakistan; several others had tickets for Jordan, Egypt, and Iraq. But Ayyad was already focused on the next mission.

His DNA was all over the letter he'd typed at his Allied Signal office and sent to the *New York Times* within hours of the bombing. The letter, resurrected later by the FBI from Ayyad's hard drive, was written on behalf of "the fifth battalion in the LIBERATION ARMY" and claimed responsibility "for the operation conducted against the WTC." Ayyad warned: "All our functional groups in the army will continue to execute our missions against military and civilian targets in and out of the U.S." As frightening as this letter was, the FBI found a second, unsent, letter in his computer: "We are the Liberation fifth battalion, again. Unfortunately, our calculations were not very accurate this time. However, we promise you the next time it will be very precise and WTC will continue to be one of our targets." It was a threat that federal prosecutors took seriously for years. Henry DePippo, who assisted in the prosecution of Ayyad, echoed him at a Senate hearing in 1998: "They warned that they would be more precise in the future and would continue to target the WTC."

When Ramzi Yousef was finally captured in Pakistan two years later, he agreed to talk to two federal agents on the plane ride to New York. After landing at Stewart Airport outside the city, he was flown by helicopter to the Metropolitan Correction Center, just blocks from his greatest triumph. As they passed the towers, an agent pointed at them: "See," he said. "You didn't get them after all." Yousef is said to have replied, "Not yet." One of the agents on the helicopter with Yousef, Lewis Schiliro, who subsequently became the head of the FBI's New York office, said, "I remember all of us discussing Yousef's return and Yousef's comment that 'Next time, we'll take the towers down.' "

Dietrich Snell, who prosecuted Yousef, says now that "it was obvious to anyone" that the World Trade Center was still a preoccupation for terrorists. His boss, U.S. Attorney Mary Jo White, added that "all of us involved in these investigations and prosecutions definitely thought the trade center was a continuing target and a very attractive one." White's views aren't post-9/11 hindsight. In a speech she delivered at the Middle East Forum a year before the attack, she mentioned the Yousef comments twice, calling them "chilling."

In fact, the FBI learned about a concrete plan to attack the World Trade Center again as early as 1995. After Abdul Murad, Yousef's partner in a plethora of lethal schemes, was captured in the Yousef bomb factory in Manila, he told Philippine investigators that the towers were among six American targets he and Yousef were preparing to attack. Yousef narrowly escaped capture himself in Manila, and the feds took his Manila plans so seriously that he was prosecuted on charges related to his and Murad's plot to blow up a dozen U.S. jumbo jets before he was tried on the '93 bombing charges. Murad, who was trained as a pilot at American flight schools, admitted to casing the WTC and providing a "structural analysis" of it to Yousef.

While only part of the Murad evidence was disclosed during Yousef's 1996 trial on the Manila counts, Murad's FBI briefings were part of the record available to Snell, White, and the rest of the Southern District counterterrorism team. It was also available to a former colleague of Giuliani's who did periodically brief him on terrorism, FBI director Louis Freeh.[1] A federal judge when the '93 bombing occurred, Freeh rushed out of his Foley Square courtroom and headed directly for the trade center. From that day on, the FBI was "obsessed" with terrorism, says Freeh, who became director that September.

Freeh's counterterrorism chief in New York, John O'Neill, who died on 9/11, also briefed Giuliani on occasion. *New York Post* reporter Murray Weiss, whose biography of O'Neill is entitled *The Man Who Warned America*, said he "reiterated since 1995 to any official in Washington who would listen" that he was sure of another homeland attack and expected the target to be "the Twin Towers again." O'Neill ran the Joint Terrorism Task Force in New York, which included up to 17 NYPD detectives, who regularly briefed their superiors. He left the FBI shortly before 9/11 and became security director at the trade center. Yet O'Neill's fears about the trade center, and Freeh's grasp of al Qaeda, never seemed to register at City Hall. Giuliani conceded that he began to broaden his knowledge of al Qaeda only after 9/11, when he got Yossef Bodansky's 1999 book, *Bin Laden: The Man Who Declared War on America* and "covered it in highlighter and notes." It warned of "spectacular terrorist strikes in Washington and/or New York," and laid out bin Laden's ties to Yousef.

Even some in Giuliani's inner circle saw the Twin Towers as a terrorist quarry. Lou Anemone, the chief of the department at the NYPD for most of Giuliani's mayoralty, says he "knew about Yousef's comments" and "knew the World Trade Center was a real continuing target," relying on O'Neill, the detectives assigned to the Joint Terrorism Task Force, and other intelligence. Starting in 1997, Anemone created a citywide security plan that ranked 1,500 buildings, transportation hubs, and shopping areas as targets. The "vulnerability list" rated the WTC as a "critical" target, the highest on Anemone's scale, as was the New York Stock Exchange, 1 Police Plaza, the Holland and Lincoln tunnels, and others. "It was very much near the top of that list, certainly in the top 20," he says.

A regular at Giuliani's weekly Wednesday 4 P.M. public safety sessions, Anemone announced his findings at a 1998 meeting. "There were very few questions. We had people in the field for six months filling out cards. When I was done briefing them on it, Rudy glazed over," said Anemone. The top uniformed officer in the department for almost six years, and a veteran of the '93 bombing himself, Anemone added, "We never had any discussion about security at the World Trade Center. We never even had a drill or exercise there.

"The threat of Islamic fundamentalism tortured me for years," the retired chief says. "I had a detective working directly for me who stayed abreast of all terrorist information. I tried to monitor it. But I felt like I was fighting an uphill battle. There was just a lack of recognition of the problem at City Hall." John Miller, a top aide for years to Police Commissioner Bill Bratton, agreed: "There were certainly no discussions regarding the possibility of a terrorist attack on the WTC at the commissioner level. I don't think anybody knew what the bombing in 1993 meant in 1994. The question of terrorism never came up."

Miller says that Bratton and City Hall "would only have discussions about specific threats or events," like the March 1994 Brooklyn Bridge shooting by a Lebanese immigrant of a school bus carrying Hasidic Jewish students. "When we learned in the middle of the night of a threat to the Israeli consulate or the Stock Exchange, he was told," Miller recalled. "We would tell him everything the FBI told us and then he'd pick up the phone and talk to Louie Freeh." But Miller does not recall any discussions "about the broader picture of what was happening with terrorism itself." There was, he concedes, "no recognition" of the need for strategic public safety assessments. Giuliani's kind of leadership—ferocious, single-minded—worked only in the here and now. To the degree he thought about planning to thwart any unspecified threats, his security upgrades were limited to the two-story buildings he lived and worked in: Gracie Mansion and City Hall.

Asked if Giuliani was accurate when he depicted himself as someone who understood the terrorist threat prior to 9/11, Miller declared, "Hello, history. Get me rewrite." The Miller/Anemone views were echoed by a host of other Giuliani insiders, from emergency management czar Jerry Hauer to Giuliani's point man on the Port Authority,

Deputy Mayor John Dyson. None of a dozen former Giuliani aides, including Deputy Mayor Fran Reiter and Department of Information Technology and Telecommunications Commissioner Ralph Balzano, could remember a single example of any expression of interest in the security of the World Trade Center on Giuliani's part. Transportation Commissioner Iris Weinshall said that "specifically on terrorist attacks, there was no sense of awareness" at the highest levels of the administration and "no discussion of the WTC." Asked if the '93 bombing ever came up, Weinshall, who later became transportation commissioner, said, "No. Isn't that odd?"

WHAT MADE IT even odder was that a civil suit brought by the victims of the 1993 bombing, proceeding slowly in state courts throughout the Giuliani years, unearthed a mountain of evidence pinpointing the towers as a target. The suit quoted two FBI officials who'd warned that the trade center was a "likely" target even before it was bombed. It cited a 1984 Port Authority report that warned: "The World Trade Center should be considered a prime target for domestic and international terrorists." The Port Authority's executive director at the time, Peter Goldmark, was so concerned that he established a counterterrorism unit, traveled to Scotland Yard, and went to the Reagan State Department to meet with counterterrorism chief Robert Oakley.

Indeed, Manhattan Supreme Court judge Stanley Sklar ultimately ruled that the plaintiffs in this suit—both families of the 1993 dead and the living wounded—had "presented proof that the Port Authority had foreseen the risk, or that the risk was foreseeable." On the basis of the authority's "actual notice of this kind of terrorist activity" prior to 1993, Sklar ordered that the case go to trial, which it did, resulting in a jury verdict in 2005 that could result in damages of up to $1.8 billion.

Ironically, the U.S. attorney throughout the '80s—when Goldmark and an assortment of consultants identified the World Trade Center as a prime target—was Giuliani. In fact, one Port Authority report at the time listed 25 downtown bombing incidents, almost all of which occurred while Giuliani was federal prosecutor, headquartered half a mile from the WTC. It cited this geographic pattern as evidence of the threat to the towers 8 years before the first attack and 16 years before the second, yet Giuliani didn't appear to understand the danger, even after 1993.

Governor Cuomo and the Port Authority certainly recognized that the trade center remained a target after the bombing. Executive director Stan Brezenoff hired a top security firm, Kroll's, and it developed a series of security recommendations for the towers. The 9/11 Commission would find that the Port Authority "spent an initial $100 million to make physical, structural, and technological improvements to the WTC, as well as to enhance its fire safety plan and reorganize and bolster its fire safety and security staffs." The commission cited a coterie of improvements: "state-of-the-art fire command posts" in each lobby, a "sophisticated, computerized fire alarm system," and "substantial enhancements" to stairwell lighting. A full-time fire safety director position was created, with supervision over a team of deputies on duty 24 hours a day. The WTC began holding fire drills twice a year. Fluorescent markings were added to lead people to the stairwells in the dark.

The changes, which came over a period of several years, would turn out to be both prescient and far short of what was needed. Vital fireproofing improvements were delayed. Recommendations that remote emergency power be installed were ignored, leaving the towers without a working public address system when terrorists struck again. Nothing was done to reinforce stairway shafts or improve exit access to the street, as '93 bombing studies urged. City Council leaders demanded emergency overrides on electronic door locks in '93, but the Port Authority moved in the opposite direction on rooftop, elevator, and other locks. None of the drills over the years involved anyone beyond fire officials, and tenants never left their own floors. The authority assigned certified fire safety directors to the sky lobbies in 1993, and quietly removed them four years later, well before the 78th-floor sky lobbies became particular scenes of chaos crying out for direction on 9/11.

Cuomo, Brezenoff, and PA deputy director Tony Shorris said that neither the new mayor nor anyone in his office ever asked about the

WTC security improvements. It's hardly surprising that a mayor who never discussed the trade center's vulnerability, the '93 bombing, or terrorism with his own key staff failed to raise the subject with state or authority officials. But it was not just a matter of hidebound indifference. Giuliani had no interest in WTC safety issues because he decided, at the very outset of his administration, to go to war with the authority that owned it. Faced with a multi-billion-dollar deficit he inherited from David Dinkins, Giuliani was looking everywhere for gap-closing cash. He quickly focused on the minuscule rents the city was collecting from the Port Authority for the two airports it had built on land leased from the city, LaGuardia and Kennedy. He'd begun fulminating about the leases as a candidate and, as soon as he took office, he resolved that he'd either dramatically push the rents up or take over the airports.

As Chief Anemone put it, "Rudy never talked about the trade center. It was the airports he was after." In fact, Anemone contends, Giuliani's eight-year, relentlessly hostile attempt to seize and privatize the airports "poisoned the well" with the authority, creating tensions so contentious that City Hall and the authority "couldn't even work together well enough" to orchestrate a "full-scale drill." The Police Department and the Office of Emergency Management never participated in a single emergency exercise there between '93 and 9/11; even the Fire Department had only one major simulated event at the site of its biggest fire ever.

GIULIANI WAS HARDLY the first New York City official to notice that it was getting the short end of the stick from the bi-state authority. New Jersey commuters enjoyed cheap, heavily subsidized transportation that city subway riders and New York suburbanites could only envy. LaGuardia and JFK looked positively neglected compared to what the Port Authority had done for Newark Airport. Anyone could see, even without the spreadsheets Giuliani was collecting from his advisers, how much more money the city could get in rent and taxes if a private manager ran the airports and a private developer owned the World Trade Center. And it was certainly understandable that the mayor who unified the city's transit and regular police was ruffled by an independent Port Authority Police Department that paid its cops better than he could pay his.

The problem with Giuliani's feud with the authority was not that he didn't have a case. It was that all he had was a case. Beyond this indictment, there was no working agenda, nor any concern about the awkward timing of his all-out assault, coming on the heels of a real, and terribly damaging, act of war.

Giuliani would virtually admit as much years later—in his April 20, 2004, private testimony before the 9/11 Commission. Asked to describe his "relationship" with the authority, he confided, "I had friction with the Port Authority re the quality of the airports, which I thought [were] substandard. Also, there was friction because the FDNY and NYPD didn't have jurisdiction in incidents at airports." As telling a summary of his eight-year war with the authority as that comment was, Giuliani also told the commission investigators: "Keep in mind, the World Trade Center was not in the City of New York." Since the Twin Towers were such an integral part of the city that they literally cast a shadow over City Hall, Giuliani was apparently referring to the fact that the authority was an island unto itself, eschewing city police and other services. But the comment inadvertently conveyed the policies of distance and dispute that were the hallmarks of his Port Authority policies, with Giuliani even briefly cutting off garbage pickup to the World Trade Center and other authority properties in 1996.

The mayor had absolutely no ability to compartmentalize, to fight with another governmental entity about one matter while working normally with it on another. As hard as it may be to believe, no one in a city administration that took office 10 scant months after the bombing ever asked any top Port Authority official how it was recovering from a terrorist hit unparalleled in U.S. history.[2] Instead, the new administration was silent about the trade center's sudden red ink, new security needs, and future financial prospects. It was as if the complex was its own city-state, rather than a vertical slice of New York with more daytime dwellers than most neighborhoods. Of course, the administration couldn't ask about the trade center if it was barely talking to the authority, and Giuliani never spoke to Robert Boyle, the Port Authority executive director during the four years preceding 9/11, and he spoke just on public occasions with George Marlin and Brezenoff, the two prior directors.

Giuliani's only public comments about the trade center between '93 and 9/11 were occasional attempts to extract larger payments in lieu of taxes from the Port Authority for the property-tax-exempt towers. When that failed and the new governor, George Pataki, made it clear that he wanted to privatize the complex, Giuliani insisted that if the World Trade Center were sold, the new owner would have to pay full city taxes. His position complicated any possible transfer and helped delay it until July 2001. With the WTC transaction on a back burner, Giuliani accelerated his pursuit of the airports—the undisputed number one profit-making division of the authority. The bizarre juxtaposition of Giuliani pushing for the privatization of public airports while erecting barriers to a public authority getting out of the real estate business passed unnoticed, though it was the reverse of reality everywhere in America.

If it was a budgetary imperative that started Rudy's war with the Port Authority, it was strange that the airport obsession only grew darker as the city's deficit turned into the largest surplus in history. If it was a preference for private-sector airport management, Giuliani certainly appeared to abandon the privatization cause when he reneged on a campaign pledge to sell the city's Off-Track Betting Corporation, a patronage plum he'd branded "the only bookie in the world that lost money." But even as these early rationales waned, Giuliani got louder, turning the bombed and embattled authority into the white whale of his administration.

THE BATTLE FOR the airports started with Candidate Giuliani. Danica Gallagher, a policy adviser to Giuliani in the 1993 campaign, says a "briefing paper" was prepared on the additional revenue the city could get from a fair airport rent or sale, helping it close a budget gap without new taxes. Gallagher met with Ron Lauder, the son of cosmetics scion Estée Lauder who'd run against Giuliani in 1989 in the

GOP primary, and Steve Savas, a privatization academic tied to the conservative and avidly pro-Giuliani Manhattan Institute think tank. They were pushing a city airport takeover and a long-term deal with private operators. And just as Giuliani bought the institute's tough-love welfare policies, he was intrigued by its privatization agenda. Savas and Lauder had also already approached Dinkins deputy Norman Steisel in a 1992 City Hall meeting, pushing the same agenda. But Dinkins wasn't keen on privatization or wars with powerful state authorities, so the initiative went nowhere. When Giuliani won in 1993, however, Savas was named to his transition team and retained as City Hall's privatization consultant.

In the weeks before Giuliani took office, Steisel and Brezenoff, old buddies who'd served at the top of the Koch administration together, just about reached an agreement on a renewal of the airport leases, which were scheduled to expire in 2015. But in the end-of-term confusion of the defeated mayor's last days, the renewal fell off the table. When Deputy Mayor John Dyson replaced Steisel as Brezenoff's city counterpart, the Port Authority remained quite willing to dramatically increase the rent, with Brezenoff offering $55 million annually and a $200 million up-front back-rent payment. But Dyson, a multimillionaire with as voracious an appetite for conflict as his mayor, was already preparing for war. He publicly demanded $100 million a year, astronomically more than the $3 million the Port Authority had paid in 1993 and considerably more than the $76 million that the Aviation Division earned altogether, including Newark Airport. Paying that rent, to say nothing of the $73 million a year that Dyson also said he wanted in additional payments in lieu of taxes for the trade center, was a fiscal impossibility.

The administration's priorities were on stark display in its reaction to two features the *Daily News* ran in early 1994. When the paper produced a two-page commemoration of the one-year anniversary of the bombing, banner-headlined IT COULD HAPPEN AGAIN, no one in City Hall had a word to say. Mixing scenes of the rebuilt World Trade Center with ominous warnings about "another spectacular event," the story relied on former police commissioner Ray Kelly, who observed,

"It's human nature for people to let their guard drop." Three weeks later, in another *News* cover story called RUNAWAY ROBBERY, Dyson branded the Port Authority the "absolute worst" airport manager in the country and painted an ugly picture of how the authority favored New Jersey. With a banner headline spurred by City Hall, the public war was on.

It couldn't have come at a worse time for the authority. The World Trade Center suffered a $32.5 million loss in 1993, though accounting adjustments turned the "extraordinary loss" into a technical gain on the final books. It wasn't until 1994, when adjustments weren't possible, that the Port Authority sank to its lowest bottom line in history. WTC profits alone plummeted from $51 million in 1992 to $8 million in 1994.[3]

The trade center was also facing $535 million in reconstruction and related improvements and a sharp projected decline in revenues. The hotel that was part of the complex, the Vista, did not reopen until November 1994, costing the authority $80 million in revenue. The top-of-the-tower extravaganza, Windows on the World, one of the highest-grossing restaurants anywhere, shut down for more than three years after the bombing, and the operators walked away. The multimillion-dollar public parking garage in the basement never reopened. Office occupancy declined, too, with high-end tenants whose leases were about to expire exploiting the uncertainty caused by the attack to drive their rents down. Other tenants simply left, paying $11 million in 1993 to buy out their leases early. A Deloitte Touche study commissioned by the Port Authority revealed that WTC security costs per square foot were over four times the average for comparable properties. Insurance costs tripled, forcing Deloitte to conclude that "the huge increase" was a consequence of the bombing.

The WTC decline made the Port Authority more dependent on the airports' revenue than ever before. Aviation was the only one of five PA divisions still producing a substantial profit. "WTC's ability to generate net income to support other businesses will be hampered by the lingering effects of the bombing," concluded a 1994 Port Authority budget office study. "Aviation's ability to generate net income sufficient to offset losses from the PA's deficit businesses is expected to become increasingly important." The same internal report listed the "outcome of rent negotiations with the City of New York" as the number one challenge for Aviation and a "critical issue" facing each of the Authority's major divisions. That's how connected the city's airport rent war and trade center improvements were.

A suddenly slimming-down Port Authority cut $100 million in 1995, including 600 staff positions and $10 million in WTC costs. Charles Maikish, the WTC director, was even cutting security services, claiming in a July 1995 memo that he'd reduced the cost of civilian guards from $20 million to $11 million while "holding the line" on proposed cuts of the uniformed Port Authority Police. He promised, just two years after the bombing, "to further scale back security costs" without inviting "unacceptable risks."

None of this duress had any impact on city policy. In fact, the $100 million rent demand that Dyson announced in early 1994 looked like a smoke screen by the end of the year. In December, Dyson suspended all lease negotiations, making it clear that the real goal was to force a fire sale to the city, which would then lease the airports to a private operator. The Lauder/Savas line had apparently prevailed. Then, in late 1995, the city took the lease dispute to arbitration. When it lost one of the key contested issues, the city's lawyers decided to just let the case sit in limbo. Giuliani's official negotiating position—which did not change for the rest of his term in office—was that the city would not discuss a lease extension on any terms, only termination and land surrender. It would not even pursue its own arbitration case.

The Giuliani hard line was a prescription for paralysis. The authority viewed the strategy of stonewalling the lease until the city got the airports as a threat to its viability that transcended Dyson's extraordinary rent demands. Resisting this challenge to its stellar moneymaker was, to the Port Authority, a matter of institutional life or death.

IT WASN'T JUST the lease talks that wound up frozen. For nearly eight years, there was no relationship between the city and the most powerful public authority in the region. Stan Brezenoff, who was in charge of the Port Authority for the first 13 months of Giuliani, George Marlin, who ran it until January 1997, and Robert Boyle, executive director until early 2001, describe the palpable tension that continued right up to 9/11. It is unclear whether Neil Levin, who was the director for four months prior to 9/11 and who died in the World Trade Center that day, was also kept at arm's length. Giuliani told the 9/11 Commission that he "normally would have spoken" to Levin after the attack, but when he couldn't, "I didn't speak to anyone from the Port Authority." Incredibly, though Giuliani was on his feet for 18 hours on 9/11, he never talked to anyone at the agency that owned the towers, even about its own substantial casualties. His antipathy to the authority was so well known that when he attended a memorial service for 9/11 Port Authority employees, he got a cool reception.[4]

Brezenoff, a liberal Democrat who'd been Ed Koch's first deputy mayor, said he "tried to talk to Rudy" about the airports, but when his $55 million-a-year offer got no response, he knew "we weren't getting a new deal of any type." Pataki appointee Marlin, the Conservative Party candidate for mayor in 1993, recalls a momentary thawing of the cold war when TWA Flight 800 went down off Long Island in 1996. He, Giuliani, and Pataki spent most of a week at Kennedy Airport meeting with victim families, doing press conferences, and even hosting a memorial. Though the Port Authority and the city worked well together in crisis mode then, Marlin said, the good feelings had no carryover. "They soon resumed the war," he recalled. "Every day, every week, it was just a constant pounding."

Boyle, a construction executive and longtime Pataki friend, says that, despite his many requests to meet, "Giuliani never spoke to me." Boyle did get an appointment with Dyson that he thought might open the door to a new city relationship. But he was disappointed: "He took me on a tour of the renovations at City Hall and sent me on my way." Boyle insisted that the Port Authority was "never at war with the city" and "only wanted to sue for peace." But, doing "nothing aggressive," said Boyle, "got us nowhere."

Intransigence was high virtue in Rudy Giuliani's City Hall, and top aides like Dyson, who'd helped jump-start the war, were licensed to accuse the Port Authority of "plundering" the city and to brand it "the fattest sacred cow in America." In an interview years later, Dyson called bullying the authority a "sport" and said he "had a good time" playing it. He blamed the breakdown in the lease negotiations on Brezenoff's unwillingness to give the city an escalator clause in the airport lease that would allow rents to increase with revenues. But Brezenoff contends that Dyson was unwilling to accept a clause that permitted rents also to drop with revenues. Obviously, these are the kind of differences routinely worked out between reasonable parties who wish to resolve them. In any event, Dyson does not dispute that the Giuliani goal quickly became to take over and privatize the airports, which, of course, foreclosed any compromise on lease terms.

No one at City Hall seemed to care about the one unmistakable by-product of this uncompromising antipathy: namely, that there were, as Brezenoff put it, "never any discussions about security concerns" or safety improvements at the towers. Marlin and Boyle had the same experience. The city's demands even threatened the financing for the security improvements planned in response to Kroll's and other studies. "Taking $150 million a year or more out would have been a heavy hit," says Boyle. "In 1994, 1995, 1996, the era Giuliani was first talking about doing it, we absolutely couldn't afford it. Even in 1997, it was still pretty tight. We couldn't have done both that and the security upgrades. Something would have had to suffer." Giuliani was forcing the Port Authority to choose between a pricey peace with him or safeguarding its terror-targeted assets.

What really enraged Brezenoff, Marlin, and Boyle, however, was the impact that Giuliani's lease assault had on the Port Authority's ability to raise money for capital construction projects. Whenever the authority needed financing, it had traditionally sold 35-year consolidated bonds, which were repaid from net revenues. But, the suspension of lease negotiations threatened its capacity to float airport project bonds that matured beyond the 2015 life of the leases. At the same time that Giuliani was demeaning the Port Authority's management of the airports, he was consciously making it more difficult for them to finance airport upgrades.

When Dyson suspended the negotiations in 1994, his top aide, Clay Lifflander, publicly warned that the Port Authority "could face trouble selling long-term bonds if would-be bondholders fear the profitable airports will not be there to pay off the debt." Lifflander urged bondholders and rating agencies to "look" at the authority's cash flow with the new understanding that its airports were "not greater than 20-year assets." At the same time, Dyson freely predicted in news accounts that the authority would come to its senses and give up the airports. Brezenoff was so upset when he read the Lifflander/Dyson comments while on vacation that he called Peter Powers, the first deputy mayor, to complain about the damage to PA bond sales. He was met with icy indifference.

Then, just to tighten the squeeze, when Dyson noticed a year later that the Port Authority was claiming, in a bond solicitation document, that negotiations with the city had only been "temporarily suspended," he wrote a letter requiring the authority to "cease describing the termination of discussions as temporary." The city, he declared, "has permanently closed any discussion of a simple lease extension and is willing to negotiate only a transfer of the airports" back to it.

While Giuliani's strategy primarily affected airport-specific bonds, it was also designed to weaken the authority's overall financing capacity. If bond buyers feared the Port Authority might lose its most lucrative holdings, that would make it tougher to sell bonds for any project. In fact, the authority's outside consultant, Deloitte, found in a 1995 study that it had been "slow in implementing" the WTC security improvements because of its money crunch. A 1996 Giuliani press release pointed out that the authority was even borrowing to "cover cash flow deficits"—to the tune of $262 million in 1995 alone.

To get around the city's refusal to sign off on 35-year bonds, Marlin says, the Port Authority "began holding the term of our new bonds to 20, and then 15, years." Since the new bonds no longer exceeded the 2015 expiration date of the lease, Marlin said, they also "no longer needed a signoff from Dyson." But the problem was not just with the issuance of new bonds. The authority also had nearly a billion dollars in outstanding bonds whose expiration date exceeded the life of the lease. It had to purchase or refund the bonds, which it started doing in 1994. This restructuring, prompted solely by Giuliani's bombast, added $41 million a year to the Port Authority's debt service. This extraordinary "Giuliani tax," as authority officials branded it, undercut operations across the board, including at the World Trade Center.

The strategy still thrilled Dyson years later: "When we started, there was 20 years to go on the lease, so they could issue 20 year bonds, then 15. The interest rates went up. Well, you can't really issue 10-year bonds, so it was getting harder and harder for them to operate. I kept thinking of Peter Pan, where the crocodile swallows the clock and the closer he gets to Captain Hook, the louder the clock ticked. It was getting harder and harder for them to finance themselves and we were getting closer and closer to taking the airports."

The crocodile, of course, eventually consumed Captain Hook.

OTHER THAN GIULIANI hardball, there was one other constant in the war for the airports: a company called BAA USA Inc. When British prime minister Margaret Thatcher tried to get the government out of the airport business in 1987, the British Airport Authority became BAA Inc. The company ran Heathrow and several other British airports, and then began branching out all over the world. Since at least 1991, the number one target of its American subsidiary, BAA USA Inc., was the airport business of the Port Authority.

Ron Lauder and Steve Savas actually brought BAA officials with them when they met with Dinkins's deputy mayor Steisel in 1992 to discuss Kennedy and LaGuardia.[5] Savas had just edited a report that included two chapters that focused glowingly on the BAA experience. The report was published by the New York State Senate's Advisory Commission on Privatization, which Lauder chaired as an appointee of the senate Republican leadership. The same thing happened in 1995, when George Marlin took over the Port Authority. Lauder asked Marlin to come to his office to discuss airport privatization. Waiting at Lauder's office were several BAA USA Inc. executives, who tried to convince him to sell the airports. Well-connected and aggressive, the company soon won a contract to manage airport facilities in Indianapolis, whose new Republican mayor was a working ally of Rudy Giuliani's.

BAA quietly became one of the driving forces behind the mayor's airport war, attaching itself to an array of Giuliani insiders. Though Giuliani was America's most recognizable mayor before 9/11, little was known outside the city about his governing style beyond his crime-busting reputation. Elected as a good-government reformer, he was widely seen by those who observed his government closely as personally incorruptible, yet quite comfortable with the lobbyists, wire-pullers, and deal makers who financed his campaigns and mined his City Hall.

For example, Randy Mastro, the deputy mayor who led the fight to drive mob companies out of the private carting business, left the administration and became the attorney for the large waste companies that replaced them. Randy Levine, another deputy mayor who went on to become president of the Yankees, was on Major League Baseball's payroll for $800,000 while he was at City Hall aiding projects for both the Mets and the Yankees owners. The wife of Rudy's chief of staff formed her own lobbying firm, and her husband helped set up appointments for her with commissioners. Giuliani's boa constrictor campaign finance committee squeezed contributions from so many people doing business with his government that he violated the city's $7,700-per-person ceiling 150 times and paid $242,930 in fines, close to a record. While the rhetoric of Jersey bias and privatization had jump-started Giuliani's airport campaign, one factor driving it over time was the myriad Giuliani ties to BAA, the special interest with the most to gain, the one that ultimately wound up as Giuliani's selection to manage the airports.

The first sign of that intertwine came in July 1995, when the city issued a request for proposals seeking a consulting firm to do "a strategic assessment" of the airports, with a focus on "the continuing role of the Port Authority." Given the mayor's already well-publicized attitude about the issues, the report's inevitable conclusions could have been quickly scratched on the back of an envelope. Instead, in January 1996, the city awarded the $600,000 contract to Wilbur Ross, the senior managing partner of Rothschild Inc., though Rothschild had acted as the financial adviser on the British privatization of BAA in the '80s. In fact, Rothschild was involved with a BAA stock deal as late as the mid-'90s.

Ross's real connection to City Hall was Dyson, a friend for 30 years. The two aging millionaires were so close that the deputy mayor was one of Ross's groomsmen at his ill-fated marriage to New York's mercurial lieutenant governor, Betsy McCaughey. Dyson acknowledged later that he "knew Ross had been the adviser to BAA" on its initial privatization, but claims he "didn't know the company would be interested" in the New York airports. If so, Dyson, who concedes he first met with BAA in 1994 or 1995, was one of the few people involved in the airport battles who was confused about its interests. BAA had all but announced in pro-privatization books and journals that it regarded JFK and LaGuardia as the linchpins of a potentially vast U.S. business.

In an interview, Ross at first said privatization "was one of the two sectors" of Rothschild he "was responsible for," listing companies and countries where he did major privatization deals, even specifying British Airways and his work for Margaret Thatcher. But when pressed about BAA's privatization, he suddenly drew a blank, neither denying he had anything to do with it nor affirming it. He said he didn't recall if Rothschild was marketing BAA stock as late as the mid-'90s, precisely when he was doing his airport work for the city, though he conceded that it was "logical" that his company might have been.

By the time Ross's selection to do the strategic assessment was announced in January 1996, the scope of work had changed. City Hall's press release said Ross would "attempt to create a picture of what the city would look like without the Port Authority." It was as if he had been hired to write a declaration of war. His brief May report did not disappoint. In addition to the predictable bashing of the authority's airport management, Ross offered the only assessment of the impact of the '93 bombing in the long archival paper trail of the Giuliani years.[6] "The World Trade Center would be considered a premium property if it were not so old, and if there had not been the 1993 bombing," he wrote. Citing vacancy rates and expenses, he put a measly $400 million price tag on the complex, an indication of just how financially damaged the Giuliani team believed it was. Instead of citing this terrorist-deflated WTC value as a cause for concern, Ross, Dyson, and Giuliani seemed to see it as a sign of weakness—evidence that this was the best time to go for the airport jugular. Ross said years later, "I don't remember any discussions about the 1993 attack" at City Hall. "But that wasn't my concern," he added.

The biggest surprise was how unabashedly the report sang the praises of a single airport management firm, BAA. It cited the Indianapolis and Pittsburgh "recent experiences," both BAA's, as the only examples that "confirm the benefits achieved internationally from public/private partnerships," even though the Indianapolis project had barely begun. "BAA's total airport operations are larger than those of the Port Authority," observed Ross in language unusual for a policy study. "In fact its revenues are as great as Port Authority's from all PA activities."

Once the Ross report was done, it took three more years for Giuliani to come up with a plan to force the issue, hamstrung no doubt by the reality that the Port Authority had a lease that wouldn't expire until 2015. Finally, in October 1999, convinced at last that the bond squeeze and other strategies would not force the authority to divest, the city put out to bid a contract unlike any it had ever awarded. It sought an airport management firm that would audit and monitor the airports until the city gained control of them. When the authority gave up the lease or 2015 arrived—whichever came first—the contractor would then automatically become the long-term operator.

While certainly not the first mayor to deliver a lucrative city asset over to a private vendor for a term that ran on for decades, he was the first to try to do it with an asset the city would not control for another 15 years. Dyson put together a selection panel to pick the winning bid and, despite Rothschild's ties to BAA, Ross was named special adviser to the panel. Ross promptly went to London and spent a day, as he described it later, "touring Heathrow and meeting with a whole group of BAA executives."

On May 4, 2000, Giuliani announced that BAA had won the bid.

The timing was a measure of how important it was to him. The announcement came a couple of days after he revealed his relationship with Judy Nathan and a couple of days before he disclosed on TV, without forewarning his wife, that he wanted a divorce. Those events were bracketed by the late-April revelation that he had prostate cancer and the mid-May declaration that he would not run against Hillary Clinton. Much of the rest of Giuliani's government came to a standstill while he was considering his medical, marital, and political options. But not BAA.

The insider appearances engulfing the deal were as evident as its extraordinary terms. The company had retained Peter Powers, Giuliani's former first deputy mayor whose relationship with Giuliani, dating back to high school, had won him the unofficial City Hall title of "First Friend." Before leaving the government in 1996, Powers had both arranged and participated in Ross's sessions with the mayor about what to do with the airports and the Port Authority. Powers's nascent company, Powers Global Strategies, never registered as a lobbyist, but instead gave "strategic advice" to clients, several of whom did city business. Seth Kaye, who produced scorching anti–Port Authority memos as a mayoral adviser and participated on the five-member panel that picked BAA, was so close to Powers that years later he joined Powers's firm.

Powers, who later became a member of the board of Ross's company, was supposed to simply advise clients like BAA, not lobby. But he reached out to the Port Authority's Boyle to set up a meeting for the company and to Marty McLaughlin, the lobbyist closest to then City Council speaker Peter Vallone. At Powers's urging, BAA retained McLaughlin to try to secure Vallone's support for the deal.[7] BAA also hired public relations powerhouse and lobbyist Howard Rubenstein, who rented office space to Powers and frequently shared clients with him.

Finally, the company retained Fischbein, Badillo, Wagner and Harding, a lobbying firm that had become the juiciest scandal of the Giuliani years. Ray Harding, the chain-smoking, ironfisted boss of the New York State Liberal Party, was a member of the firm and the undisputed heavyweight of influence peddling in the Giuliani years.

Harding had given Giuliani his party's endorsement in 1989 and 1993, and it eventually provided his margin of victory. Saddled with a quarter-million-dollar tax lien when Giuliani took office, Harding joined the Fischbein firm, became an instant millionaire, insinuated both of his sons at the highest levels of the administration, and watched his firm's client list climb to a new stratosphere of perceived clout.

One of Harding's sons, Russell, ran the city's Housing Development Corporation—a position that allowed him to loot the city for $400,000. Russell later went to jail on fraud and Internet child porn charges. He was recorded on a Web chat room at the time of the BAA deal as telling a friend, "The mayor is always helping my dad.... He had my dad go over some contracts a while back about a company taking over the airports here ... so not only in the long run will the mayor profit from it ... but my dad already has with the contract.... They both have a wash each other's back thing going." When his chat room partner expressed confusion, Harding replied, "It makes perfect sense for the line of work my dad is in ... as well as my job.... We all help the mayor and he helps us."[8]

Dyson was closely attached to Harding, having virtually singlehandedly bankrolled the Liberal Party in the '80s, when he ran on its line for the U.S. Senate. Fran Reiter, another Liberal leader who was deputy mayor in Giuliani's first term, says that "Ray was Dyson's rabbi" and helped him get his City Hall title. "Yes, BAA had Ray for a while," Dyson acknowledged. "Ray's entitled." Harding also gave his ballot line for governor in 1998 to Ross's wife. Completing this incestuous circle, Rick Fischbein, the managing partner of Harding's firm, was also a business partner of Ross's, and Rubenstein represented Dyson's upstate family winery. The Fischbein firm, the Dyson family, Powers, and Rubenstein combined to raise and contribute hundreds of thousands of dollars for Giuliani campaigns over the years, with Dyson and family dumping $20,000 into Giuliani's federal political action committee as late as 2004.

BAA and the city signed a letter of intent in December 2000 and hammered out what Dyson said was an "inch thick contract" in 2001. While the terms were never fully disclosed, one sweetener did become public: an estimated $200 to $300 million breakup penalty that any future mayor would have to pay BAA if he tried to cancel the deal. Dyson said, "Rudy and BAA's chairman met personally several times," which, of course, was several times more than he met with the head of the public agency that actually ran the airports. BAA brass flew from London to meet Rudy and "we went over there several times," Dyson recalled.

In the midst of this invisible intrigue, Giuliani climbed atop the shrillest soapbox he'd ever been on in all the years of Port Authority wars. When a major snowstorm hit the city on December 30, 2000—after the letter of intent with BAA was signed but before it was announced—the mayor went on a helicopter ride over Kennedy Airport and took some photos himself. "Not a plow in sight," he charged at a January press conference. The authority responded that it had 440 people and 77 pieces of equipment cleaning up the foot of snow. "This is like, you know, what was that Soviet news agency, Tass, right?" Giuliani replied. Feigning a Russian accent, he said, "Da airport ees clear. Der ees no snow on da airport." A *Newsday* editorial observed that the city was witnessing "a four-alarm outbreak of Rudyrage."

The editorials didn't slow him down. Giuliani claimed he was going to sue to retake the airports, using the snowed-in runways as his rationale. He said his photos showed that the Port Authority was using only two of its four runways because of the snow, unaware that if it used all four, the planes would collide. Then, on the heels of a week of abuse, he announced BAA's letter of intent. But the Giuliani onslaught came to a public halt when Vallone insisted that the BAA deal had to go through the long-winded community review processes that had been the death of many far more worthy initiatives. Nonetheless, Giuliani reluctantly began it. Whatever was driving this deal, the only force strong enough to stop it was September 11.

When the planes hit the Twin Towers, John Dyson was seen rushing out of City Hall with Rudy's massive Rolodex in his arms, trying to find the mayor.[9] "If there was no 9/11," he insists, "we would have worked out a compromise with BAA. That's what I thought and that's what BAA thought." But 9/11 did happen, and Giuliani never raised the question of BAA again.

On the final tally sheet of the costs of this chronic conflict, Rudy managed not only to cumulatively impose hundreds of millions of dollars in damaging new bond expenses on the Port Authority, he also deprived his city of hundreds of millions in remarkably higher rents. For example, in 1994 and 1995, when the city was still gripped by daunting deficits, it got a combined $24 million in airport rent, $86 million less than the Brezenoff offer. In Mike Bloomberg's first State of the City speech in 2002, he announced a "revitalized relationship" with the Port Authority, eventually renewing the airport lease until 2050.

GIULIANI HAD ALSO missed countless opportunities to make the largest buildings in the city safer. Dyson, for example, says he has "no idea what happened" to State Senator Roy Goodman's bill, which sought to place the trade center under city building and fire code jurisdiction. Brezenoff, Marlin, and Boyle say that no one from the Giuliani administration ever raised the issue with them, even though Goodman kept introducing his bill every year until 9/11, joined eventually by the Democratic assemblywoman who represented much of downtown. Dyson could hardly claim ignorance—his aide penning the most acrid airport memos was Seth Kaye, the former Goodman assistant who'd first brought the exemption bill to the attention of the Giuliani campaign.

Even without the Goodman bill, the city still had the leverage of the 1993 memos of understanding that Brezenoff signed with the Dinkins buildings and fire departments right after the bombing. The memos did not grant the city the enforcement power it had over the rest of the city's skyscraper stock. But the agreement with the Department of Buildings did guarantee that the Port Authority would adhere to the city code and "resolve expeditiously" any disagreements the department might have with its conformance. It also required the authority to maintain a code compliance file "for any project" at the World Trade Center. City officials "may at any time request a copy of any Project file," the agreement provided.

The memo of understanding was signed the day after Giuliani was elected in November 1993, granting new powers to a new administration. Yet a review of thousands of pages of Department of Buildings documents related to the towers did not uncover a single department request to review a project file over the nearly eight Giuliani years.[10] Not only did the department do nothing to use this agreement to tackle the safety vulnerabilities of the complex, it actually agreed in 1995 to water it down. A supplement signed by Giuliani's buildings commissioner, Joel Miele, eliminated the requirement that the authority "thoroughly review and examine all plans" to determine whether tenant improvements conformed to the code. Instead, it allowed the tenants to retain an architect to certify compliance.

Giuliani fire commissioner Howard Safir signed a similar 1995 amendment to the Fire Department agreement with the authority. But since the Fire Department's original memo of understanding went further than the building department's—requiring the Port Authority to submit fire safety system changes for "review and approval"—Safir's amendment actually gutted the 1993 agreement. It eliminated any requirement that the drawings and specifications be submitted to the Bureau of Fire Prevention and surrendered the bureau's right to approve them. Instead, architects retained by either the authority or a tenant could simply certify compliance. Like the Department of Buildings, the Fire Department could view the authority's project file whenever it wanted, and it, too, never did.[11]

What was particularly shocking about the watering down of the memos of understanding was that the Port Authority never asked for any of the changes. The preamble to the February 1995 amendment flatly states that "the FDNY has requested, and the Port Authority is agreeable" to eliminating the fire safety requirements. Signing the amendment days before the second anniversary of the bombing, Safir volunteered to surrender oversight powers that the Dinkins administration had fought to secure.

The Department of Buildings amendment also seemed to be the city's idea. Under Miele, the department had begun a deregulation plan that allowed architects retained by private builders to self-certify their projects. Deregulation was a City Hall initiative, favored by the real estate and construction lobbies, and the modification of the memo of understanding extended it to the Port Authority. While Giuliani ally Goodman declared during his postbombing hearings in 1993 that public safety matters at the World Trade Center "can't be a matter of self-policing," that's precisely what Giuliani's administration made it. The overall Giuliani self-certification program wound up, over the years, denounced by district attorneys as ripe for abuse and criticized in independent audits. When the National Institute of Standards and Technology looked at it as part of its 9/11 probe, it urged an end to any form of "self-approval for code enforcement."

Yet the Giuliani administration rushed in 1995 to extend this laissez-faire approach to an authority the mayor routinely denounced, sacrificing powers that were granted only because the 1993 bombing revealed how necessary they were. In fact, under the terms of the Department of Buildings amendment, the authority had more self-certification latitude than deregulated private developers. From 15 to 25 percent of the private, self-certified plans were spot-checked by city inspectors to determine whether they met code. But because of this amendment, none of the Port Authority plans would be checked. And the deregulation program barred architects who designed private projects from also certifying a project's code compliance, while architects on authority projects could do so.

In the spirit of these self-circumscribed powers, Giuliani's administration rolled over on two of the most important safety issues at the WTC—fireproofing and stairwells.

Within the Port Authority, alarms had been going off for years before 9/11 about the WTC's puny half-inch fireproofing. A 2000 PA study determined that it was useless to try to patch missing areas of it because the gaps were so large it was "more effective to replace it." A review ten years earlier found sections where the insulation was "very sparse to a quarter inch" and trusses "adjacent to the outside walls" that were altogether "devoid of fireproofing." After the '93 bombing, the authority had begun taking measurements of the thickness of the insulation on North Tower floors. It determined that it needed to triple the fireproofing to an inch and a half so that the two-hour protection requirement of the code "can be achieved." The fact that the towers weren't in compliance with city standards was echoed in 1997 by a consultant, Rolf Jensen & Associates, that studied WTC life safety systems for the Port Authority and conceded that "the towers' structural steel fireproofing fall somewhat short of the required" two-hour standard.

A third internal report in 1999 included an assessment of the fireproofing in a comprehensive review of all trade center code issues. That memo emphasized the three categories of "non-conforming code items" at the World Trade Center first identified by the safety consultant. The Port Authority would do nothing about Category A items, said the memo, while Category B deficiencies such as fireproofing were being "remedied," and plans to correct Category C shortcomings would be developed "in the near future." The National Institute of Standards and Technology concluded that "these reports were available for review by the city," though it did not indicate that anyone ever asked to see them. Department of Buildings records do not contain any assessment or objections it submitted to the authority about any of these categories of violations, even those that the PA planned to do nothing to correct.

The Port Authority started to slowly retrofit the floors in 1995, but waited until 1999 to adopt a formal policy requiring that all vacated floors be reinsulated before a new tenant could move in. By 9/11, only 31 of the 220 floors had been upgraded. The authority's upgrade peaked in 1998, when 11 floors were retrofitted, and slowed to a crawl by 2000, when only one was.[12] This multi-million-dollar code compliance project was precisely what the Department of Buildings and the Fire Department were supposed to oversee under the 1993 memos of understanding. But there is no record in either department's files that it ever bothered to review any of the 31 floor upgrades, much less object to the authority's snail-paced effort to do anything about its disturbing noncompliance.

Under the terms of the Fire Department's original memo of understanding, the retrofitting was, at least arguably, a fire safety system project that would have required approval before it was begun. Had that agreement still been in place, the department could have insisted that it had the power to review the studies and object to the timetable, since fireproofing directly related to the length of time that it had to extinguish a fire in the towers. But Safir had relinquished the department's fire safety review rights, and the authority just proceeded with its foot-dragging upgrade.

Regardless of the effect all this had on 9/11, the fireproofing chronology reveals how negligent City Hall was for years about protecting WTC occupants. While the post-9/11 investigative record is unclear, there are indications that the refireproofing delays may have had deadly consequences that day. The National Institute of Standards and Technology concluded that "the fireproofing played a key role in the structural response to the fires" and the collapse of the towers. It found that the North Tower fire floors were reinsulated and stood for just under two hours, almost precisely the protection an inch and a half was supposed to provide. But only one of the six impact floors in the South Tower was upgraded, with the rest still insulated by the half-inch fireproofing installed during construction. Tests have determined that a half inch provides three-quarters of an hour of protection, and the South Tower stood for 56 minutes.

Nevertheless, NIST concluded in 2005 that the effect was not significant because the impact of the planes "dislodged" whatever fireproofing was there. This appeared to conflict with its June 2004 report, which attributed the insulation breakdown to both causes—the planes that dislodged the fireproofing and the public agencies that failed to ensure that adequate initial or upgraded protection was installed.

A NIST engineer, Dr. Kuldeep Prasad, compared the thickness and dislodging factors at a 2004 session, and clearly saw both as causes of the collapse. He conceded that the South Tower floor trusses "heated more quickly" than the North Tower trusses, a conclusion ostensibly tied to the South Tower's relatively puny insulation. But he then contended that "the removal of the fireproofing during impact was more important than the differences in the initial thickness of the fireproofing." While NIST may rate dislodging ahead of thickness as factors in the collapse, this conclusion still makes the scant fireproofing on the South Tower floors, which the city and authority took no action to correct for six years, a cause of the tower's relatively rapid demise. There are also indications that if the South Tower had been upgraded, less of its minimal fireproofing might have been dislodged by impact. The primer paint used in the original fireproofing reduced its adhesive strength by a third to a half, making the South Tower floors more vulnerable to dislodging impact than the unprimed, retrofitted floors.

Of course, fireproofing buys time, even if no one can determine whether an upgrade of the South Tower might have extended its life for 10, 15, or 30 minutes. A sizable number of above-impact South Tower occupants had apparently discovered the passable Stairway A and may have been minutes away from safety when the tower collapsed. Fire chief Orio Palmer, marshal Ron Bucca, and other firefighters were in the impact area, rescuing 18 people stuck in an elevator, when the crash came. Palmer had dispatched other firefighters to help several injured people he'd found on the 70th floor, and indicated in his last radio messages that he would soon get above the fire. One 9/11 family group, the Skyscraper Safety Campaign, has wondered out loud whether upgraded fireproofing in the South Tower might have added precious minutes and saved dozens of lives—questions that have no definitive answer.

In addition to the city's negligence in forcing quicker code compliance on the fireproofing, its war with the authority may have contributed to the delays in the improvements that *were* happening. As much as it has been praised for the safety improvements it made after the 1993 bombing, the Port Authority took forever to do many of them. The stairwell emergency lighting and illuminated exit signs, for example, weren't finished until 2000, despite all the outrage about the dark descent during the '93 evacuation. And the fireproofing, of course, never was completed. It's hard to judge how many of the delays were due to the WTC post-1993 budget crunch, and how many were simply the result of its well-known bureaucratic lethargy. But Boyle says that if the authority ever agreed to the city's financial demands, it would have delayed these already plodding programs even more, perhaps leaving the North Tower floors as unprotected as the South Tower ones were that day.

The Giuliani administration also abetted the authority's resistance to stairway improvements. There were too few stairways—three instead of the code-required four in each tower. Most of them emptied into the mezzanine instead of the code-required street access. The stairways were also too close together and protected only by partition walls—invitations to destruction that were nonetheless possibly code compliant. "Had there been a minimum structural integrity requirement to satisfy normal building and fire safety considerations" in the building core, the National Institute of Standards and Technology concluded that "the damage to the stairways, especially at the floors of impact, may have been less extensive." Had there been a fourth stairwell per tower, NIST found that it "could have remained passable, allowing evacuation by an unknown number of additional occupants from above the floors of impact."

Instead of examining these stairway issues after the '93 bombing, when the passageways out of the towers emerged as a crucial public concern, Giuliani's Department of Buildings (DOB) actively participated in the authority's disregard of the code and other evacuation concerns. When new management began the refurbishing of Windows on the World in 1994, the question of a fourth stairway, for example, rose to the fore. A fourth exit is required under code when a "public assembly area" designed for over a thousand people is operating in a building, and the two-story restaurant in the North Tower, as well as the Observation Deck in the South Tower, met that standard. In December 1994, with the original memo of understanding still in effect, the authority met with the top brass of the buildings department to try to seek approval of the restaurant plan, which provided for so-called "areas of refuge" on the restaurant floors rather than a fourth stairway. But records of the meeting indicate that, to the authority's surprise, the DOB officials didn't even bother to raise the issue of the fourth stairway. Instead, they quickly agreed to misinterpret an exception to the code, which gives a pass to dining facilities that only serve building occupants. The exemption exists because employee cafeterias don't result in an increase in the building population, while restaurants and observation decks do.[13]

The buildings department did much the same on the mezzanine exits. The 1997 safety consultant report and 1999 internal authority memo listed the failure of the exit stairs to discharge to a street as a code violation. In 1993 and again on 9/11, the police and fire departments had to establish circuitous routes out of the buildings because the stairs dumped occupants in the mezzanines. The Department of Buildings had looked the other way in 1975 when the towers opened, but the 1993 evacuation experience, the memo of understanding, and the new code conformance policies put the issue on the table again. The PA position was a joke. It equated the plaza outside the towers and the concourse underneath them with streets. And the buildings department permitted this unambiguous violation. When the 1999 memo listed the mezzanine exits as a Category A nonconforming item it would not correct, the Giuliani administration did nothing.

There were 188 people in Windows on the World on September 11 and a few more in the Observation Deck area. Under the city's own code, neither should have been allowed to reopen after 1993 without a fourth stairway, which, NIST found, could have saved lives. Of course, shutting down Windows, where Giuliani hosted his second inaugural breakfast in January 1998, was always unlikely, as was forcing a redesign of the mezzanine exits. But these two code violations, as well as the questionable lack of stairway separation, would have given a vigilant Department of Buildings the leverage to force the Port Authority to at least reinforce the living room–like walls that protected the shafts. Contemplating the too-few stairways too close together and out-of-code exits, the city could have pressed the authority to do what was most achievable—strengthen the integrity of the walls protecting them.

But that would have required will and imagination, both of which were in short supply at the buildings department. Like the Fire Department, the Department of Buildings was just another patronage backwater in the Giuliani era. The first Giuliani commissioner resigned in a month, shoved out by revelations that his asbestos removal company was under federal indictment in Boston. The politically wired Miele ran a small Queens engineering firm whose clients included associates of mob boss John Gotti. A partner in the Ray Harding law firm championed the appointment of another acting commissioner close to him. The Republican City Council leader was involved in Department of Buildings scandals that led to the prosecution of a onetime department official close to him. The mayor was so uninterested in independent professional judgment at the department that he dumped another commissioner who suggested that Yankee Stadium didn't have to be replaced.

In fact, Giuliani was so uncomfortable with tougher code requirements that he introduced only one significant amendment during his eight years—in sharp contrast with his predecessors. That change, extending sprinkler requirements to residential buildings, occurred only after seven people, including three firefighters, died in two fires within a matter of days. He had actually opposed a City Council sprinkler bill a year earlier, and squashed a similar Fire Department proposal in 1994. The Department of Buildings' acquiescence to the Port Authority was symptomatic of this Giuliani indifference.

So was a report Miele and Fire Commissioner Safir issued in 1995. Rudolph Rinaldi and Carlos Rivera, the buildings and fire commissioners under Dinkins, had started the study, naming a 40-member World Trade Center Task Force that included the Port Authority, the real estate industry, architects, engineers, and top city officials. The two commissioners had issued their own report a month after the 1993 bombing, calling for legislation to bring the World Trade Center under the city's code jurisdiction. They bemoaned the fact that "code compliance at the WTC has been dealt with by every fire commissioner and Chief of the Department over the last 25 years," with fire officials reduced to relying "on persuasion to gain compliance."

The task force they created was charged with reviewing "all fire safety areas of the code" to determine whether there were "advances that would make buildings safer," as well as to try to adapt the code "to take terrorist actions into consideration." At a minimum, Rivera and Rinaldi said the study would explore new requirements "to prevent a local explosion from incapacitating entire buildings," such as mandating "remote locations for emergency generators."

The group continued working under Giuliani's commissioners, but in the end, the Miele/Safir report did not recommend a single material code change. The task force said it "could not reach unanimity" on the emergency power issues that Rinaldi and Rivera said it would deal with "at a minimum." Acknowledging that the failure of the emergency public address system led to injuries during the 1993 evacuation, it nonetheless found that the code requirements for communications systems were "beneficial and adequate." It considered a host of issues that mattered on 9/11 and did nothing about any of them—the evacuation of the disabled, mandatory full-time fire safety directors, restrictions on locking devices to exit and roof access doors, and a reinforced fire tower stairwell.

The report did urge the new Office of Emergency Management that Giuliani was in the process of creating to develop an "emergency commercial broadcast protocol to provide emergency communication to occupants in high-rise office buildings." But when the OEM started a few months later, it was never even given a copy of the report.

In the immediate aftermath of the '93 bombing, everyone from Mayor Dinkins to the fire and buildings commissioners to Senator Goodman to the leadership of the City Council was calling for an overhaul of the code and jurisdictional control over the Port Authority. No one in the Giuliani administration heard that call. Instead of fighting the good fight to protect the tens of thousands of New Yorkers and others who worked in or visited the targeted World Trade Center every day, Rudy Giuliani dedicated his two terms to a seedy airport power grab masquerading as civic virtue.

THE PORT AUTHORITY Police, a particular foil for Giuliani, offered the best evidence on 9/11 of just how intertwined the authority's far-flung airport, trade center, and other assets were. Three Port Authority cops assigned to the airports rushed in to help at the towers and died; only four of the 37 who died that day actually worked at the World Trade Center. The others, including Police Superintendent Fred Morrone, sped there from bus terminals, tunnels, bridges, headquarters, and the emergency services unit. The mother of one dead Kennedy Airport cop, 16-year officer George Howard, made international news when she approached President Bush on September 15, 2001, at a Javits Center gathering and handed him her son's police shield. Howard had dashed to the WTC from the airport without anyone ordering him to.

In the middle of a tsunami of Giuliani charges against the authority just eight months before 9/11, Bernie Kerik got into a spit fight with Morrone and Port Authority Inspector Anthony Infante, who supervised the Kennedy Airport police and, like Morrone, died on 9/11. Kerik challenged the comparative competence of its force. Giuliani himself contended that if the city policed the airports, "you'd have significantly less organized crime influence," accusing the Port Authority cops of not cleaning it up. None of this was forgotten among the authority rank and file, or its institutional leaders, when history came full circle on 9/11.

On that day, Rudy Giuliani was still in court against the authority, as well as pushing his BAA takeover, after years of ignoring the many ways he could have helped make its World Trade Center safer. He'd held his press conference, mocking the Port Authority as an American politburo just a few months earlier. Only when he saw its prize towers in ruins did Giuliani pull back. In sorrow and in anger, the long-besieged and hardly faultless Port Authority, with 84 of its employees and contractors dead, was finally, on September 12, at peace with the city it served.

CHAPTER 5
WATER MAIN WARS

AFTER NEARLY 35 years in the New York City Fire Department and two as its highest-ranking uniformed officer, Anthony Fusco was finally ready to retire in 1993. But he had one last job he wanted to do. When terrorists bombed the World Trade Center in February of that year, Fusco had managed the operation—the largest incident in his department's 128-year history. He was proud of what the firefighters had done. "We didn't lose another person after the six in the initial blast even though the smoke conditions were horrible," he remembered. "We had to reach over walls. We had many rescue situations. It was very fulfilling. It was the most rewarding incident I'd been to."

As a legacy, Fusco produced two detailed accounts of the successes and shortcomings of that operation. They were published at the end of 1993, shortly before Rudy Giuliani took office. He wanted the reports, which included articles by him and others in the department's top echelon, to shape preparation for future terror attacks, so he and his colleagues pulled no punches. The question "Was there something we could say that would improve operations?" was what drove him. The list of issues he detailed eight years before the terrorists would strike again was chilling in its prophetic aura.

- *Lack of cooperation between fire and police.* As incident commander in 1993, Fusco was supposed to be in charge. But he had to move his command post to a tent to get away from low-level representatives of a dozen other agencies who clustered around, getting in everyone's way, as he recalled it. Meanwhile, the high-level decision makers who were supposed to represent each agency at the command post were nowhere to be found, even after Fusco sent a chief to try to bring them in. The police insisted on staying "at their own field headquarters," he said. His recommended solution was "routine interagency drills" with police, firefighters, and others preparing for high-rise fire and other major events in a unified way.

- *The defend-in-place strategy.* Instructing occupants of big buildings to stay put until firefighters reached them had been the foundation of the Fire Department's strategy for decades, the absolute core of the city's high-rise protocol. But Fusco said it "did not exist" at the towers. Everyone who could evacuate themselves did it as fast as they could, ignoring advice to the contrary. Since everyone could see that was the right choice, a flexible new strategy for high-rise occupants was necessary.

- *Communication.* Blasting Fire Department radios as "overloaded and ineffective," Fusco said "in many cases, runners were sent by a sector commander to communicate with the incident commander." He said the department was "currently in the process" of developing a new interagency radio system.

- *The building and fire codes.* Disturbed by the ways the World Trade Center "did not comply" with the city codes, Fusco urged new legislation to make sure the city had the power to enforce its rules. Fusco and the rest of the brass complained about elevator shafts, mezzanine exits, power backup, and thousands of locked internal doors. "If the fire department is to fight the fire," he concluded, "the fire department should enforce the codes."

- *The elevators.* Fusco called rescuing the hundreds of people trapped in elevators "the most difficult part of this operation." He agonized over the 72 schoolchildren locked in an elevator for five hours, until a firefighter found them and started to take them out one at a time, only to have the elevator suddenly start moving because elevator mechanics were unaware of the ongoing rescue. In mega high-rise complexes like the World Trade Center, with as many as 230 elevators, a real plan for coordinated search and rescue, aided by building elevator mechanics, was underscored as a priority.

- *An emergency operations center.* A deputy Emergency Medical Services chief, Steven Kuhr, coauthored an article contending that the city needed a center to "coordinate with the command post" and oversee Fire Department deployment, 9-1-1, and other operational needs that "cannot be addressed on site during a disaster." Fusco's reports also identified the handling of disabled occupants and use of outside engineers for advice on building stability as two other areas requiring better coordination.

Fusco and his colleagues had seen something in 1993 that cried out for changing the way cities thought about handling worst-case fires in enormous skyscrapers. They weren't the only ones. The National Fire Protection Association and the Building Officials and Code Administrators International sent a team to the site immediately, and they produced a 59-page study focused, for the first time, on "mega high-rise fires." Prior to the incident, no one treated megastructures any differently than other high-rises, said the report, but it was now time to ask if that should change. Urging a rethinking of the defend-in-place strategy, the staid industry groups also concluded that megabuildings "are not designed to be totally evacuated in an emergency" and called for a code and tactical revolution.

As sobering as the entire analysis was, the hot-button item highlighted by the 1993 bombing was the war of the badges, particularly triggered by the police helicopter rescues on the roof of a tower. The hostility was so intense during the 1993 evacuation that Alan Reiss of the Port Authority "observed a confrontation between the FDNY and the NYPD that almost deteriorated into physical violence."[1] Reiss attributed the incident to the fact that the rooftop rescues were done without consulting the fire chiefs. Fusco wrote that the unauthorized landings "put lives in danger" and "caused friction." The dispute would rankle for years to come.

The Fire Chiefs Association attacked the airlift as "sheer grandstanding, a cheap publicity stunt done at the expense of public safety." Police Commissioner Ray Kelly took the polar opposite view. He testified before a March 1993 congressional hearing that at least 30 people were evacuated to safety, including a pregnant woman who was flown to a hospital and gave birth immediately. Kelly's spokeswoman fired back at the chiefs' letter, calling it "absolutely ludicrous," and charging that the chiefs were engaging in a "petty turf battle" that did "a disservice to the people of the city." The commanding officer of the aviation division, William Wilkens, crowed, "There were many surprised and exhausted firefighters who, after climbing up 80 flights, were met by fresh police rescue teams coming down the stairs."

Fusco tried to steer the controversy in a constructive direction by pushing for joint aviation rescue drills that included police and fire rooftop operations. There had been no such exercises for years. "It's just human nature," Fusco said later. "People who know each other have a much better chance of cooperating in emergencies. The more a stranger you are to each other, the more you're going to stay within your unit and do whatever you can. We felt that one of the big problems in 1993 was that the training together was somewhat neglected." A City Hall meeting to discuss helicopter protocols was held in 1993, including the mayor's office, police, fire, and other agencies, Fusco recalled. Then both departments eventually published their own, separate, rule changes.

The Fire Department's regulations said rooftop rescues could be undertaken only at the direction of the Fire Department incident commander, and it wanted the police to let fire personnel do the work. The revised police manual provided that a fire chief would be airlifted to the roof if landings were planned, but also described circumstances when rescues would be attempted without FDNY involvement. The dispute—which required integrated training and drilling to resolve—was still way up in the air when the Dinkins administration departed at the end of 1993.

But it wasn't just helicopters and separate command posts that fed the competitive animosity. In 1993, police chief Lou Anemone recalls, there was "a Mexican stand-off over the bodies." He says the police set up a temporary morgue in the nearby Vista hotel and planned to bring the bodies discovered in the basement there, but the Fire Department wanted to put them in ambulances. When a lieutenant informed him of the dispute, Anemone wanted to know—why ambulances?

"He told me he thought it had something to do with the cameras, which were right by the ambulances," Anemone contends. "So I went to the chiefs in the basement and we got into a disagreement over who was in command at that point of the incident. I said: 'You don't have to be Sherlock Holmes to figure out this is a crime scene. Are you shitting me? The fire is out.' I insisted the bodies had to go to the morgue, that we were not sending them to five different hospitals. So finally we agreed that two of my guys and two of their guys would carry each body out, and that I and the chief would lead the way up and out. When we got up to that point where you had to turn either toward the ambulances or toward the Vista morgue, we were pulling in both directions. But we got them to the morgue. They were disappointed there were no cameras."

This was the stunning state of emergency planning when Rudy Giuliani took office. Fusco anticipated the future so well that he closed one of his articles with a reference to the June bombing plot. "It didn't take long for the specter of an even larger disaster to be raised," he wrote warily. "In June, four months after the WTC attack, police arrested a group of suspected terrorist conspirators in Queens." Then he listed their targets, some of which would have involved consuming high-rise fires. He called for an "awareness of the need for preplanning and command organization beyond the scale" of anything the Fire Department and the city had ever before envisioned.

"The bombing, to me, was a violation on the soil of the United States," said Fusco, who volunteered for the service as soon as he turned 18 so he could fight in the Korean War. "It hadn't happened before. If they thought they were successful, even to a small degree, I thought they would probably say, 'Well, we'll do more in this area.' I thought the World Trade Center would be a particular target. Yes, definitely, I thought that. Just like the civil disturbances of the 60's, I viewed terrorist activities as the next major issue facing the fire department."

But the new administration wasn't listening to these warnings. When it came to emergencies, it had its own, peculiar, priorities.

TWENTY-ONE DAYS into Rudy Giuliani's first term at City Hall, a water main broke in Brooklyn. It happened at 1:30 A.M. and Giuliani found out at 5:30 A.M., watching TV news at Gracie Mansion. The gusher in middle-class Carroll Gardens turned a Brooklyn street "into a raging river," said the *Daily News*. The resulting explosion at Gracie Mansion led, eventually, to the creation of the Office of Emergency Management. The 24-hour-mayor was outraged that he was out of the loop for four hours and that reporters knew about a city emergency before he did. He got to the scene by 7 A.M., just as the rising half-million gallons of floodwater caused a street to collapse, breaching a second main and creating a half-block-long crater. Soon the river was a sheet of ice. The Department of Environmental Protection had to locate 60 valves beneath eight inches of ice to shut the main down.

A subhead in the *Daily News* story captured what would become an earmark of Giuliani governance. It simply said "Rudy on scene." At one point, said the *News*, he directed rescue workers who were removing a 94-year-old man trapped in his house. With 340 homes affected, some caked in mud, Giuliani went back and forth to the site for days, even hosting a community meeting there. He instantly threw his brand-new government into high gear about both fixing the burst main and making sure he would never again have to learn from TV that something bad was happening in his city.

First Deputy Mayor Peter Powers had already become a student of the city's response practices. Powers had been Giuliani's top operative in the transition with the Dinkins administration, and he had raised questions with Dinkins deputy mayor Norman Steisel about emergency management. Steisel said Powers indicated that they "wanted to have an emergency agency of their own and understood the public relations value to the mayor." Powers told him that "the mayor has to be involved" and that he "never wanted to have a situation where the mayor doesn't know what's going on," a not-too-subtle reference to the out-of-the-loop Dinkins of the '93 bombing and other major incidents. Steisel said Powers had questions about how to staff such an agency and was acutely aware that the "public perception was that the police department really ran emergencies," a perception Giuliani seemed determined to change. "I knew when Peter had an intense interest and he did in this issue," said Steisel. "But was it terrorism that interested him? Or did the threat to the World Trade Center? No. He was interested in gaining mayoral control of emergencies."

So the first official communication in Giuliani City Hall that is labeled "Emergency Response Coordination" is a memo from Powers to the mayor dated January 24, 1994. "As we have discussed," Powers begins, "the Brooklyn water main break in Carroll Gardens has highlighted that the City's response in emergency situations should be improved." The areas Powers says "should be addressed" included "24-hour City Hall emergency response coordination" and "the need to clearly define the notification process, particularly Mayoral notification." He was also concerned about making sure that "the City Hall press office coordinated the media response."

There actually already was an Office of Emergency Management. It was run by a police commander and it had not covered itself in glory during the water main break. A representative didn't get to the scene until 6:30 A.M., and even then went to a nearby tunnel entrance. So Powers's memo said there was a need "to recognize that other city agencies may be more adept than the police in coordinating non-enforcement emergencies." This was the earliest indication that the new administration was rethinking the longtime city plan of allowing the Police Department to oversee interagency management of all city emergencies.

Powers reviewed two possible approaches: a 50- to 60-member Emergency Response Team, which, he noted, Giuliani had suggested, or a 5- to 7-member Emergency Response Unit of "highly qualified individuals." In either case, staff from other agencies would be assigned to City Hall on rotations so that someone would be on duty at all times. Their responsibilities would include "monitoring all police desk calls, determining who should be contacted, and responding to an emergency scene, where appropriate, with the understanding of all involved that they represent you." Suddenly refocused on fighting a succession of snowstorms, Giuliani didn't respond until February 15. In a handwritten note to Powers, he decided to "train 20 people to do this to start with—no overtime." He wanted it going by that weekend, and it was. By Friday the 18th, Powers had named a full team from four agencies, none of them police or fire, and dubbed it "the emergency notification and response system at City Hall." He appointed Bill Motherway, the risk manager from the Office of Operations who'd assessed the water main response, to head it.

The whole operation was rushed together so quickly that few involved understood just what they were supposed to do. In early March, Powers asked the Emergency Response Team officer on duty about an emergency situation and, as Motherway put it in a memo, was "rather upset to discover that, not only was the individual unsure of the incident, but unsure of his actual purpose." Motherway said it was "imperative" that the team leaders from other agencies "ensure that all assigned to the ERT understand the procedures." Any failure "to do so will probably result in us splitting the nine weekly shifts ourselves," warned Motherway, who called that possibility "a fate worse than death."

Motherway could not talk about the ERT in an interview years later without breaking out in gales of laughter. "I usually stayed in Peter Powers' office" all night, he said. "They insisted that ERT had to be staffed by commissioners, deputy commissioners or assistant commissioners. They didn't even want directors; they wanted seasoned people bright enough to make a decision. You'd work a full day at your agency, then do the ERT shift from 7 p.m. to 7 a.m. I was mayor one night a week. Our responsibility was to find out what happened, then consider should we notify the mayor or make a decision on it. It was Rudy's first foray into trying to take control of the whole emergency thing. He wanted to know what was going on at all times. It's not like we walked around in gear and saved people. We kept a big logbook. Someone, I'm not sure who, looked at it."

The initial list of procedures forwarded from Powers to Giuliani fit on a page. The pivotal instruction was to be prepared to answer questions from Powers or Giuliani about any incident. Archived notes from an ERT protocol meeting months later show a hand-drawn chart with information passing along arrow lines from Police Operations to the ERT member to "RWG" [Giuliani] and Powers. The scrawl on the side added: "What does RWG want to know about? When to wake him up?" The incident sheets listed nine major forms of emergency that the group was to track, including "major structural fires, explosions and hazardous material incidents," quickly moving the team's interests beyond the water main fiascoes that prompted it.

Powers always saw this as merely a bridge to the larger agency he'd discussed with Steisel. The subject title for his May 12, 1994, memo to Giuliani was "Mayor's Office of Emergency Management." He said he'd reviewed the city's current police-run office and surveyed emergency operations in other cities, noting pointedly that in Los Angeles, Giuliani ally and Republican mayor Richard Riordan was "directly involved in the coordination and delivery of emergency response services in every major incident, something not done in prior mayoral administrations." Powers analyzed the pros and cons of keeping OEM under police supervision, contending that the Police Department operates "most effectively at enforcement-related incidents and tends to be less effective at non-enforcement related incidents." Comparing the status quo with the "transfer of emergency operations to the mayor's office," he zeroed in on the obvious advantage, noting that the new director and office would "have your authority at emergency incidents."

Donna Lynne, the director of the mayor's operations office, responded to Powers's proposed transfer in a detailed August 26 memo of opposition. The subject of terrorism never came up at either end of this 10-page exchange. It just wasn't part of the debate. Instead, the dispute focused on emergencies that were not terrorist acts, namely "non-enforcement events," such as storms or water main breaks. One reason Lynne was asked to comment was that her unit played a key role in such events. Powers's clear rationale for creating OEM was to improve the handling of these incidents. He faulted the Police Department for "making decisions" outside its expertise "without consulting other more experienced or informed agencies," specifying the Fire Department and the press office.

Allied with Anemone, who was by then the highest-ranking uniformed official in the Police Department, Lynne resisted the change, but she believed that Powers and Giuliani had already made up their minds. "By August, the decision was made that everything was going to move from the police to under the mayor," recalled Bill Motherway, who was the mayor's liaison to the police emergency management office under Dinkins and Giuliani.

Finally, in December 1994, a firebombing near City Hall closed the question. An unemployed computer programmer carried the bomb onto a train in an extortion plot against the Transit Authority. It exploded prematurely, scalding the bomber and injuring 49 people. Giuliani celebrated the city's multifaceted response at a press conference in the immediate aftermath. A subway cop saw the charred bomber arrive at a Brooklyn station and held him on suspicion. Another cop on the train rescued many people. An Emergency Medical Services technician rushed to the scene and helped "a stream of victims staggering from the smoky subway to daylight above," the *News* reported. The Fire Department was immediately at the scene and effective. "The end result," Giuliani bragged, "is that an incident that could have been very, very destabilizing turns into one that helps to stabilize the city and helps to convince people that the city can respond effectively."

But Jerry Hauer, who ultimately became the new Office of Emergency Management's first director, says Giuliani and Powers told him later that the mayor was privately seething. "He talked to guys from the fire department, the police department and EMS and couldn't get a patient count" that he needed for his press conference, said Hauer.

Steven Kuhr, the Emergency Medical Services official who would become Hauer's top deputy, said that separate command posts were stretched down Broadway and that the mayor could see that "nobody was talking to each other," even about something as basic as injuries.

"Giuliani felt that it was critical for a mayor to ensure that emergencies were well run," Hauer said in his private testimony to the 9/11 Commission.[2] "There were two incidents that led him to create OEM. First, there was the firebombing in the subway. The mayor was unable to get the full story on the event and felt that coordination between the emergency medical service, fire and police was disjointed. The second event was a water main break in Brooklyn. It was a huge street collapse and the mayor heard about it on TV. No one called him. That's what led the mayor to set up OEM." Later, deposed in a civil suit involving the collapse of 7 World Trade Center, Hauer listed the same events, emphasizing the defective patient information. "He wanted good information. He wanted fast information. And he wanted an agency that reported to him," Hauer testified.

Donna Lynne, who had overseen the Emergency Response Team, was happy to give up all of her office's emergency responsibilities, which sometimes had her rushing out to beaches to watch waves and ordering sandbags in anticipation of a hurricane. In fact, it was hurricanes, snowstorms, and water mains that absolutely, in her view, led to OEM. She couldn't recall terrorism ever even coming up in the City Hall discussions that led to the creation of the new agency.

In January 1995, a brief design for the new agency was completed, and quietly, in February, the budget office approved $200,000 to fund it. The search for an executive director began, but it dragged on, unsuccessfully, for months beyond the deadline. One applicant, Jonathan Best, who as director of emergency management in Bridgeport, Connecticut, handled the collapse of two 13-story buildings, recalled the selection process as "light" and unfocused. A former Emergency Medical Services supervisor in New York, Best said the job posting in the *Times* was "a little tiny, blind ad that didn't even say City of New York on it. It just had a P.O. Box number. It was almost as if it were a secret that they were hiring somebody."

Twelve city officials attended his interview, including the Fire Department's special operations chief Ray Downey and a police captain, and everyone was making jokes. "They didn't know what they were looking for," says Best. "They didn't know what an emergency manager really was." Was the subject of terrorism raised? "Terrorism was not mentioned. I was pretty surprised at that." What about the '93 bombing? "I don't believe it ever came up." Was he ready to discuss terrorism? "I prepared absolutely. I went over in my mind how I would respond to certain questions." But nothing like that surfaced. They gave him one scenario and asked how he would respond; it was, predictably, a hurricane. "The police guy joked that I had outlined their exact plan. That's when I thought I had done well."

Hauer didn't respond to the blind ad. A former deputy director of Emergency Medical Services in New York, he had gone on to become the director of emergency management for Indiana. He knew Downey from his New York days, and when they met at a search and rescue event, they got talking about the OEM job. By then, the search was stymied. Hauer was a heavyweight in the field, and Downey told him, "Get me your resume." Downey kept calling him, and Hauer wanted to come to town to see his mother anyway, so he agreed to an interview. It turned out to be the first of five, a prelude to the one with Giuliani. The only time he remembers the subject of terrorism coming up throughout the marathon was when the city executives discussed a sarin gas drill it had done shortly after a deadly, cult-orchestrated attack in the Tokyo subways in the spring of 1995. "They just did the drill. It turned out badly," Hauer remembered.

He eventually graduated to a City Hall session with Denny Young and other brass. They asked him to stay over and meet with Powers at seven o'clock the next morning. In a three-cigarette walk around Hauer's hotel, Powers probed his planning background and thinking. "He talked about the problems between police and fire," citing examples but not any related to the '93 bombing. Hauer even met with Giuliani's wife, Donna Hanover, who early on played a pivotal role in key decisions. Then he went back to Indiana and, a couple of weeks later, was asked to return to meet with Giuliani. "I was very impressed. We had good chemistry. He wanted an independent OEM that worked for him and did strategic planning and interagency coordination."

But when Hauer was offered the job, he initially turned it down. "The pay and status was at a deputy commissioner level," he said. "I told them the office would fail if the director was in a lower-level position, not on a par with the police and fire commissioners. He had to be a peer at the same salary and with direct access to the mayor. They had to come around to my way of thinking about it. They soon said okay and I was hired at the same salary and as a member of the cabinet."

When Hauer's appointment was announced in January 1996, he, Police Commissioner Bill Bratton, Fire Commissioner Howard Safir, and Giuliani posed for pictures in the mayor's office, and he went over to the police and fire departments for separate meetings with the commissioners and their top aides. The mayor actually cited the Tokyo attack as a reason to improve the city's preparation for increasingly difficult disasters, though Hauer said he didn't recall Giuliani ever mentioning the World Trade Center bombing to him before or after his hiring.

Hauer started in February, precisely a year after the hunt for a director started. He had already managed to redefine the job. While Giuliani told him in the interview that he wanted better coordination between the uniformed services, Hauer said, "I wouldn't say calming the police-fire rivalry was a high priority." Hauer raised that issue to a new level with his status and salary demands; the office acquired a potential authority not previously considered. He understood, even if City Hall didn't, that coordination was more than a bromide, and that doing it would take teeth.

City Hall had not given him any particular outline of how OEM was supposed to be organized. So Hauer began drafting his own plan, as well as the mayor's executive order, which empowered OEM to coordinate the city's response to all incidents requiring a multiagency response and listed a dozen categories, starting with "severe weather" and ending with "acts of terrorism." The executive order also put OEM squarely in command of the Police Department's emergency center, spelling out that Hauer would decide when it was used and "have primary responsibility for its operation" during an emergency. It was the first time an outside agency was, ever granted control over an operation inside police headquarters. Hauer's office, however, hardly had resources commensurate with its potential authority. To keep the head count down in the mayor's office, he was told he could only hire staff on the payroll of other agencies. "They didn't want to create a whole new agency," he recalls, disturbed that "no other agency was run like that."

Hauer's hand was strengthened, however, by a resurgence of attention to the debacle of the city's sarin gas drill. Just days before the official launch of his agency, the U.S. Senate Permanent Subcommittee on Investigations released excerpts from the city's confidential 1995 sarin drill after-action report at a well-covered Washington hearing. It found that the subway drill resulted in the theoretical deaths of most of the first hundred responders, who rushed in "totally unprepared." In an echo of the '93 bombing analysis by Fusco and others, the report also called radio communications "abysmal," concluding that interagency communications were impossible, since the agencies "operated on different frequencies." The Police Department's frozen zone and command post at the drill were configured for bombs, not gas, which meant that most of the personnel and vehicles there were "in the path of escaping gas." Only the Fire Department's Hazardous Material Unit performed capably, the report concluded.

Hauer said Giuliani had only "mentioned the drill in passing" during their initial talks, and, since it occurred three months after Giuliani decided to establish OEM, he knew it had nothing to do with OEM's creation. But Hauer had a strong background in bio/chem and the well-publicized release of the sarin gas report gave him leverage to push for new protocols and to put pressure on the police to accept them.

Indeed, putting pressure on the Police Department had long been part of the unacknowledged politics of the campaign that led to the creation of OEM. Hauer was quite willing years later to say that the Fire Department was a big factor pushing for OEM: "Safir was trying to screw Bratton and the police department." Asked in a civil suit what "the scope" of the new position he was offered was, Hauer testified: "to develop an agency that the mayor was pulling out of the Police Department and evolve it into a real emergency management agency."

Safir equated a Police Department deduction with a Fire Department addition. He vowed at his first public meeting with the firefighters union to retake control of emergencies from the police, a promise that brought cheers from the only major union to endorse Giuliani. Motherway says he participated in meeting with Safir as early as March 1994 "about transferring emergency management from the police department to the mayor's office." Richie Sheirer, who was promoted to deputy fire commissioner by Safir, claims that he convinced Safir to push Giuliani for "a more level playing field, with emergency management in the mayor's office." Safir "brought that to Mayor Giuliani," said Sheirer, and Giuliani "looked it over" and created OEM.

Giuliani and Safir were already on the same page—both wanted to reposition the Fire Department as the city's prime rescue agency. With the dramatic decline in fires nationally and locally, a 12,000-person FDNY was unsustainable, especially in budget-slashing times, unless Giuliani could come up with something else for it to do. So the mayor decided early in his administration to take Emergency Medical Services away from the Health and Hospitals Corporation and deliver it to the firefighters. Giuliani also leaned toward moving more emergencies under Fire Department control, and that would become Hauer's approach at OEM, especially on hazardous materials, or hazmat, incidents.

Safir's push, however, was all about turf. It was not a serious policy assessment, based on the recommendations from the '93 bombing experience. In fact, Safir appeared barely aware of them. Anthony Fusco stayed on as chief of the department for Safir's first six months, but the new commissioner never discussed the bombing with him. "Safir and I had a few discussions, but I can't recall him ever involving me in anything like that," he said later. "I guess there were ongoing projects at the time, but I can't recall him sitting down and discussing the World Trade Center, no." Safir never talked to him about OEM or terrorism, either.

Safir was not unaware of what Fusco had written, however. Unbeknownst to the retiring firefighter, Safir did allude to the Fusco reports in a letter to Peter Powers in February 1994. "Many findings" from the 1993 "post-operational review have caused us to rethink certain facets of our operations," he wrote. But Safir wasn't referring to the defend-in-place strategy, or the radios, or any of the major issues Fusco had raised. His only concern was that the department had to use reserve vehicles to fight the 1993 fire and that the reserves "lacked air packs, masks and other essential items." He wanted $300,000 to stock "the reserve apparatus."

While Donna Lynne thought the water main and other nonenforcement emergencies were the driving force that led to OEM, others theorized that Safir was. "He had a much better relationship with Rudy than we did," said Lou Anemone, the uniformed embodiment of the Police Department. "Safir would gin Rudy up. He wanted OEM as a way to get around us. OEM would put the fire department in a better position to control some emergency scenes."

When the police ran OEM, the fire chiefs and other top officials would have to make their way to the eighth floor of 1 Police Plaza during emergencies, a trek Anemone was sure the Fire Department brass "hated." Giuliani apparently didn't like it much, either. Bratton's top aide, John Miller, the former NBC and ABC reporter who's now the principal spokesman for the FBI, painted the picture: "OEM was a little room off to the side of the operations section, but big events came to life there, like snowstorms and blizzards. The mayor sought to take that over." After the first big snowstorm of his administration, Miller said, Giuliani "chafed" at having to "trot over to our office where the cameras were. It drove him nuts. The police commissioner was in charge in a crisis; why was the mayor subservient to the police department? So he hired Jerry Hauer and created a mayor's office of emergency management, and that way, the next time there's a snowstorm, the fire department and the police department were subservient to him."

Bratton put it more diplomatically: "The mayor recognized early on he could better coordinate between agencies if control rested with him. So there was a perceived need, based on watching agencies battle each other. And it also fit into his well-known penchant for absolute control."

Hauer thinks Miller and Bratton are overstating what he sees as just one element driving the campaign for OEM. "Electeds always want to be shown as in charge. There were a lot of rationales for OEM and the visibility of the mayor might have been one. I don't think you can look at it as just an in-charge issue. I'd rank that lower than others. It might have been a subtext." Actually, Hauer's agenda for his briefing with the mayor on March 19, 1996, the day the executive order was signed, lists "the visibility of the mayor's office/the mayor" as one of the three functions of the new office. Deputy Mayor Fran Reiter, who was close enough to watch these machinations but was not a prime actor on police/fire issues, said, "I can't tell you whether terror was part of the equation. It was more a question of who's in charge. The historical struggle between police and fire was on the mayor's mind. He loved cops, but he did not have a utopian view of how they functioned. They were in control of their thing, the city's response to emergencies. He put in place a structure that would be immune to internecine warfare. That's where OEM came from—civilian control over response. I can't say any of it came from a recognition of terrorism."

It was a water main break. Conflicting patient counts from a madman's firebombing. Blizzards and the threat of hurricanes. A need to know first. Safir's turf-war whispers. A grand stage. A good-government attempt at stronger coordination. Realism about police efficiency. OEM was a mix of all of these intertwined motives—temper tantrums, reform, friendship, union allies, ego, and control. The one thing it was not was a prophetic vision, as Giuliani and others have portrayed it. Terrorism was as far down on the list of motives as it was on the list of anticipated incidents in the executive order. The '93 bombing came in behind a foot of snow. Instead of foresight about an apocalypse, OEM was an example of hindsight about a 30-foot water pipe.

* * *

The Police Department was the strongest public bureaucracy in New York, a city unto itself. It reacted to any perceived challenge to its power as fiercely as it reacted to a 10-13, the radio call when an officer is in distress. To Jerry Hauer, who was trying to make the NYPD a team player in an arena it had long controlled, steely resistance to a unified emergency plan was no surprise. What was unexpected was that Rudy Giuliani let the department undercut Hauer's supposed mandate to re-organize the city's response preparations time and time again.

They could get away with just about anything. Detectives tailed him secretly. The intelligence division wouldn't give him any information, although he had the highest security clearance of anyone in the administration. Even after his office was officially put in charge of the command center at police headquarters, the department refused to tell him if its backup generators worked. It disapproved 16-hour tours for the drivers that drove him and key staff to emergencies, routinely told OEM's 24-hour Watch Command to call 9-1-1 about emergencies, just like any citizen. It boycotted a hazmat/terror drill and delayed the assignment of selected police staff to OEM. As late as the eve of the high-alert millennium, the grand debut of Giuliani's command center, the Police Department was still defying City Hall orders and preparing to run its own operation out of the supposedly mothballed emergency center at 1 Police Plaza.

If the good-government purpose of OEM was to get police and fire on the same emergency response page, Hauer, instead, wound up with a book-length manuscript of contentious memos and dueling drafts of response plans. But what wore him down, and finally forced him to leave, wasn't so much the infighting. It was the waffling at City Hall, the "lack of support" he got. He was so well-regarded that he wound up with the top bio/chem job in the Bush administration, so his departure in 2000 was a noticeable loss when catastrophe struck a year later.

When Hauer's appointment was announced in January 1996, he sat down with Commissioner Bratton, anticipating combat. But Bratton was laser-focused on street crime and unthreatened by a new emergency office. In his own book, *Turnaround*, which is principally a postmortem on his New York years, Bratton doesn't mention his own or Hauer's emergency management offices, nor does he refer to terrorism, the '93 bombing, hurricane threats, water mains, or any other crisis response question. They were not high on the agenda, Miller acknowledges. Bratton and Hauer connected instantly, and as it turned out, all they had was an instant to connect. "Bratton couldn't have been more gracious. I got no resistance from Bratton," Hauer recalls.

As successful as Bratton had been, he was already on Giuliani's hit list when Hauer arrived. The trouble started with a *Time* magazine cover story, also in January, picturing Bratton in a trench coat by a squad car and making him the poster boy for crime reduction. By March 1996, as Hauer was settling in at his operations office, the Giuliani team was anonymously carving Bratton up in the press, planting stories about his personnel moves, six-figure book deal, nightlife, and junkets. On March 26, Bratton quit, later acknowledging that he was driven from office.

Two days after Bratton resigned, Howard Safir was named commissioner. Safir had law enforcement credentials as a drug enforcement agent and an associate director of the U.S. Marshals Service, but all of it was collateral experience compared with the background of a career urban cop like Bratton, who understood policing in his gut. Giuliani, who had his corporation counsel probing the propriety of Bratton's recent book deal, was strangely undisturbed by the fact that Safir had unsuccessfully marketed his own book for years and been successfully sued by a ghostwriter he had hired. In the days when memoirs were supposed to bear some resemblance to truth, Safir advertised himself as a marine, though his entire active-duty career consisted of two, six-week, platoon leadership sessions at Quantico. He dropped out in the middle of the second session, just as he'd flunked out of law school. His claimed role in the capture of an Asian drug lord was also mostly farce, but his dozen rejection letters from publishing houses were quite real. ("I thought about all the events I had participated in and realized that there were few major crimes, disasters or government conspiracies that I has [sic] not had some contact with," Safir declared in the book proposal.)

Safir's prime initiative as fire commissioner had been his effort to shut down the street alarm box system, and his campaign to sell the plan was as big a loser as the one for his book. He rallied support for the alarm box initiative by telling a reporter: "Will there ever come a time when a person won't be able to get to a phone and report an emergency and result in a death? The answer is yes." When it came to the ability to spin a quote or charm a reporter, Safir was Bratton's mirror opposite. He once told his federal staff before a visit by senate investigators, "If any of you have dirty laundry, you'd better keep it to yourself," and the quote wound up in a newspaper. It was *Newsday* columnist Ellis Henican, however, who put his finger on Safir's greatest selling point at City Hall: "Say what you will about Howard Safir, but Rudolph Giuliani will never again experience the indignity of being overshadowed by a police commissioner more charismatic than he." Safir's six-foot-three frame, steel gray hair, stone face, and aura of authority was perfect for his on-camera role, which was to stand solemnly next to the mayor when crime dominated the news.

What few admirers or detractors knew about Safir was how adeptly, despite his size 14 shoes, he could pirouette. Jerry Hauer was about to find out. Though Safir and sidekick Sheirer, who came with the new commissioner to the Police Department, had helped invent OEM as a vehicle for Fire Department ascendancy, they switched sides as soon as they switched headquarters.

Safir was sworn in on April 15 in the most elaborate ceremony on the steps of City Hall for any appointee in history. By July 2, the agenda for Hauer's monthly meeting with the mayor celebrated OEM's early successes but then listed a few issues: "Success has its price. Not getting the support from Howard. They cannot micro-manage us. We get little information in areas we have asked for and agreed we should get. They refuse to allow us to get involved. Howard has simply not given us the backing." Steering OEM already in the direction of bio/chem planning, Hauer specifically cited "terrorist threat" information as one of his frustrations.

By September 25, 1996, Hauer's notes indicated that Safir was criticizing OEM for "trying to get to the media" with emergency information rather than "the mayor first." Hauer's response to Safir's critique concluded: "Howard's veiled threat about stonewalling us is what we see happening now." The notes also included the astonishing revelation that six months after Safir became police commissioner, "Howard asked me who in the Police Department he could talk with on terrorism."[3]

In January 1997, Sheirer, a new top deputy in the Police Department, took the budding war to another level. Sheirer became Safir's point man in leading the opposition to Hauer's first major directive, a draft plan for how to coordinate a hazardous materials or chemical event. Hauer's plan made the Fire Department the incident commander. That was precisely what Safir had called for when he ran the Fire Department, but ensconced now at 1 Police Plaza, the new police commissioner discovered all kinds of hazards in the idea. Attack-dog Sheirer said in a letter that the plan missed "the mark in terms of understanding of the NYPD as an organization, its culture, and its expertise." An attachment to the Sheirer letter called the plan "an attempt to pick the pockets of the Police Department," by excluding it from chemical identification, medical triage, decontamination, "and informing the press." Under no circumstances, the review concluded, "should the Police Department relinquish authority or responsibility to maintain public order to any other city agency."

Sheirer also took some shots at the department he'd worked at for the previous two and a half decades. Objecting to the requirement that the police and other agencies consult with the Fire Department on protective gear at the scene, the Sheirer attachment said: "This assumes the FD member responding has the necessary expertise, which may not be the case." The same language was used about evacuation. "Does the Fire Department have the necessary expertise to determine the evacuation areas?" His memo also challenged the guideline that permitted the Fire Department to decide when an incident reached "a solely investigation stage" and, thus, became police business. It warned that giving the Fire Department such authority "will not only create animosity, but will lead to the infamous 'battle of the badges.' "

The bottom line was that Sheirer's package of assessments branded the plan "potentially dangerous to citizens of our city who might be hazardous material victims," calling it "fool hearty [sic] and unacceptable." It also trashed OEM, even recommending the elimination of the term "On Scene Coordinator" that OEM had given itself, and charging that the plan "does not reflect any input given by the PD during this planning process." Sheirer continued pummeling the plan with a March memo, and the Police Department then produced its own 29-page draft. The police charged that all Hauer was doing was "following what's generally done throughout the country"—a strange argument—and one that Hauer answered by saying there wasn't another city in America where police run hazmat events.

The battle never abated. As late as December 3, 1998, the Police Department formally informed Hauer that it would not participate in a hazmat drill scheduled for December 7 because participating "would result in our condoning this dangerous, misguided and improper intrusion" into a matter that belonged under police jurisdiction. Safir was out to kill OEM—the best idea he'd ever brought to the Giuliani administration. In his proprietary mind, every addition for the Fire Department was now a subtraction from police power.

The Police Department waged the same guerrilla warfare against Hauer's second most important initiative, a 1998 plan for biological incidents that also put the Fire Department in charge. This time Anemone led the paper trail of opposition, making the same "crime scene" presumption that the police offered on the chemical plan. The detailed critique contended that the first agency to respond should contact the Police Department's 9-1-1, not OEM, and repeatedly attacked any "operational role" OEM planned to assume. The police argued that the incident commander "in a terrorist incident" of any type, including the anthrax example that was part of Hauer's protocol, "is either the FBI under federal law or, since terrorism is a crime, the Police Department." That argument was the department's rationale for going it alone whenever terrorism was suspected.

Echoing the '93 bombing tug-of-war over World Trade Center bodies, the Police Department was particularly upset that the Fire Department got special coffins "known as containment vessels." Handwritten notes in the police assessment of the Hauer plan complained: "PD never got its coffins," and, then, simply "circumvent the protocol." In Hauer's efforts to try to satisfy competing interests, the protocol called for teams of two Fire Department Hazmat, two police Emergency Services, and two Department of Health personnel to simultaneously enter the room where a biologic agent was believed present. The Police Department was offended that it was a firefighter who was selected, as the police memo put it, "to peek into the room and take digital pictures" before entry.

The war even infected the planning for the millennium, which was widely seen by city and federal officials as a moment of maximum terrorist threat. On June 26, 1999, just as Hauer and Giuliani's long-awaited command center at 7 WTC was coming on line, the Police Department sent out letters to a wide variety of agencies and groups announcing that its own emergency center would be open for business December 29 through January 3, 2000. It was a direct, defiant challenge to Hauer and, indirectly, Giuliani. When City Hall got wind of the Safir end run, Joe Lhota, the mayor's first deputy, issued an unmistakable October 14 order aimed directly at Safir but addressed to all agency heads. "In an effort to consolidate the response" to millennium disruptions, OEM's command center "will be the single, central Emergency Operations Center." Incredibly, five days later, Ray Powers, deputy chief of the Police Department's operations division, sent a follow-up letter to those who got the June one, complaining that the police had yet to receive a reply to their request that organizations "provide a representative for our command and control center." Even after a terrorist conspiracy targeting New York was uncovered by the FBI days before the millennium, the department went ahead with its own command center operation, fragmenting the response that night.

Hauer announced his resignation shortly before the millennium and left soon after, dismayed at how few of his detailed response plans had actually been put in place. "Every agency involved in the hazmat/ chemical and biological plans had to sign off on the protocols," he said, "but Safir never did." Instead, Hauer got "page after page of critical notes" from Sheirer, along with word that "the police commissioner didn't agree with the protocols and that they were not going to adhere to them."

Anemone, who felt OEM was simply "giving more to the Fire Department than should be," agreed that the police should fight the rules for responding to biological or chemical emergencies. And he remembered that "the mayor never came down on one side or the other. I don't think he ever made a decision that the police department would be in charge or the fire department. Safir wouldn't sign off, he was dealing with Denny Young and Rudy and telling them the Police Department had issues, and Rudy didn't make him sign. Jerry came back to us with more plans and we said 'Jerry, you've got to deal with the issues.' For such a decisive guy, Rudy did nothing." That meant that no approved rules for these kinds of incidents were in place, even at the millennium, a moment of maximum threat.

Phil McArdle, a member of a Fire Department hazmat unit and a particularly aggressive leader of the firefighters union, wrote a letter in October 1997 about the police acting "in direct contradiction" of hazmat protocols. "Nobody in the City of New York can control these individuals," he charged. "No matter what documents are written, by the mayor, by the Office of Emergency Management; the Police will selectively choose what they will comply with. Is this not anarchy?" McArdle wasn't referring to the unsigned regulations; he was more frustrated by police stonewalling of what he called "a mayoral directive sent to all agencies."

Hauer wrote, and Giuliani theoretically sanctioned, the directive that obsessed McArdle. Developed in early 1997, the four-page document called "Direction and Control of Emergencies" contained an organizational chart, or matrix, that listed which department was in charge of every anticipated multiagency response. The Fire Department had hazardous materials, nuclear, radioactive release, and chemical crises. Biological was undetermined, with the caveat that OEM would "task agencies based on the type of event." (That language preceded the 1998 bio plan, which put the Fire Department in charge.) The introduction to the matrix called it "preliminary" and said the assignments were "suggested for the purpose of fostering a dialogue among the PD, FD and OEM." The designations were "intended to present questions" to be resolved in joint planning. At the same time, the Hauer-prepared document invoked the executive order's powers and said it directed "OEM to define and ensure participation of all agencies" with emergency roles. The assignments were described as a result of that mayoral mandate.

Hauer framed the matrix and put it up on his office wall. He acted as if it had the force of law. After the Police Department refused to sign the protocols, it became all Hauer had. At first, the department contended that it had no idea what Hauer was talking about, explicitly stating it had never so much as seen a copy. Then, in a September 19, 1997, letter, Anemone acknowledged that the department had at last seen this tentative assignment list. Without conceding anything, Anemone said simply, "According to OEM, this matrix is now city policy."

Ray Kelly, who got into his own battles in the Bloomberg era and won control over hazmat incidents, researched the history of the issue carefully. "The order came up and isn't signed by Giuliani," Kelly insists. "No one can tell you when it was issued; it's undated. The mayor was ambiguous about issuing the order." Hauer insists that he still has his signed and framed copy, but concedes that the mayor's lack of support for the protocols that implemented the matrix cast such a cloud over it that no one took it seriously.

Perhaps the best evidence of the ambiguous authority of Hauer's organizational chart is that Giuliani was forced, four years later, to issue one in July 2001 with precisely the same title. By then, Richie Sheirer was in charge at OEM, a fox-in-the-chicken-coop promotion that stunned and deflated Hauer. This second matrix was definitely signed by Rudy Giuliani and was in effect on 9/11. Instead of resolving the disputed questions of terrorist chemical, biological, or nuclear events, it passed the buck, saying: "NYPD or FDNY." In a footnote, the two-page directive said: "The nature of this type of event is such that the Incident Commander will shift as the event evolves." It laid out three possible transfers of authority at the same terrorist event—

from police to fire and back to police, with OEM resolving conflicts. The Fire Department did get control of hazmat incidents that were not terrorist events, though how anyone would determine which ones were acts of terror and which ones weren't at the outset of any event was a mystery.

If this was supposed to be the definitive Giuliani command and control order, it proved, after nearly eight years in office, that the mayor had neither command nor control. By then, Bernard Kerik and Tom Von Essen were running police and fire. Faced with a choice between uniforms, with his friends at the helms of competing empires and Sheirer a blank slate, all Giuliani could produce was mush. The 9/11 Commission's John Farmer said Giuliani "wouldn't have had to issue the 2001 order if command and control was already in place," confirming how uncertain the 1997 assignment of responsibility was. "There was an effort to improve communication and the relationship between the departments," said Farmer. "But I don't think they made inroads."

The Police Department wasn't satisfied to merely frustrate Hauer; soon it targeted him. "I had to make my own contacts at the CIA and FBI," he said, "because I was not getting the threat information the Police Department was getting. The intelligence division was told not to talk to OEM." While the police wouldn't give Hauer any information, they were quite eager to get some on him. In 1998, prior to the hazmat drill the police boycotted, police detectives tailed Hauer to a Fire Department prep session at Randall's Island in upper Manhattan. "They took pictures of me working with the fire department. Howard brought the pictures to a joint meeting with the mayor," recalled Hauer. "Howard said I was giving an unfair advantage to the fire department by helping them prepare for this drill. The photos had me standing with the fire chiefs. I told the mayor I would help any agency. I was tailed like a common criminal and they put it in front of the mayor. My driver told me I was often tailed." Hauer saw the confrontation as a test of Giuliani's support: "I was pretty disappointed. He just looked at me. I thought he would have been more outraged at having a commissioner followed."

Hauer recalled a similar second session, this time in 1999, shortly before the millennium. Hauer, Safir, and Von Essen were meeting with Giuliani, and Hauer told the mayor that the Fire Department was excluded from Police Department millennium security briefings. "Rudy looked at Safir and asked how much the fire department was involved. Howard said he got them involved in the planning. I said: 'Don't lie to the mayor.' Howard was shaking. I said: 'Tommy, answer him.' And Von Essen said: 'Well, we really haven't been involved.' "

It baffled Hauer how Safir could get away with so much. The police commissioner had become a public scandal, all the invisible emergency control issues aside. The press revealed that he had assigned two dozen cops to work at the new Police Museum, where his wife was board chair. He had used on-duty detectives to chauffeur his daughter's wedding guests. When his wife was involved in a fender bender, he'd filed suit, demanding $250,000 for loss of sex. The police union took an unprecedented vote of no confidence in him. When unarmed Amadou Diallo was killed in a 41-shot police attack and protests filled 1 Police Plaza, the commissioner skipped a City Council hearing on it because of "a scheduling conflict," only to be caught on camera standing next to Helen Hunt at the Oscars. Even Giuliani's ethics board wound up slapping him on the wrist for taking a $7,000 "freebie" to Hollywood, after it was revealed his benefactor did minor business with the Police Department and after Safir finally ponied up for the junket.

Hauer was long gone by the time Safir resigned in August 2000, hit by prostate cancer at the same time as Giuliani. On his last day in office, Giuliani called his friend of 25 years "the greatest commissioner in the history of the city." Safir was as modest, telling a large audience at a police promotion ceremony that his administration was "the most successful in the history of the NYPD." Giuliani immediately installed an even closer and more subservient friend than Safir, Bernie Kerik.

But the mayor's protracted unwillingness to stand up to Safir or the department, even in service of emergency response goals he publicly proclaimed, has remained a secret to this day. Jerry Hauer's sworn testimony about Giuliani's spinelessness in a civil suit years later left a roomful of lawyers, on both sides of a case involving the collapse of

7 WTC, in utter disbelief. The lawyer for Con Ed, whose substation was destroyed in the collapse, asked Hauer whether the operations center at 1 Police Plaza, which Hauer oversaw from 1996 to 1999, had a backup emergency generator that worked:

HAUER: I don't know if it was functioning.

QUESTION: You were the head of OEM?

HAUER: Yes. The police would not give us a straight answer. They told us it was working. They never tested it on load. The Police Department barely allowed us in that command center. We got nothing but resistance from the Police Department.

QUESTION: Am I correct, sir, that you as director of OEM never satisfied yourself as to whether or not there was an operative emergency generating system to protect the command center?

HAUER: The Police Department would never give us a correct answer.

QUESTION: Did you ever go to your friend Rudy Giuliani and ask him to get an answer for you?

HAUER: I did.

QUESTION: Did he?

HAUER: No.

QUESTION: Why not?

HAUER: The Police Department was resistant—

QUESTION: Wouldn't give him an answer either?

HAUER: We never got a straight answer—

QUESTION: Are you telling me, sir, that you, as director of OEM, went to the Mayor of the City of New York and asked him to find out whether there was an operative emergency generating system for the command center of OEM, and he was unable to do so? Is that your testimony?

HAUER: I am telling you that I told the mayor that we could not determine whether the generator at 1 Police Plaza was functional.

QUESTION: If you thought an emergency generating system was critical to the functioning of OEM, why didn't you do something to find out whether you had one, no matter what that took?

HAUER: Because we were battling to try and survive, because the Police Department fought us right and left. They did not want OEM to survive. They undermined us every step of the way.

It was a side of tough-guy Rudy few in New York knew. Cowed by a department that worked for him, a rollover. He would not take the Police Department on, especially after he drove Bratton out. Under Safir and Kerik, it was really his department. It surrounded him with the largest detail in history; he lived inside its cocoon. He called police headquarters early every morning and got the crime numbers that were making him a national hero; the overnight tallies were usually better than breakfast. He could still see four of his uncles in their uniforms. His federal law enforcement career made him, in his own mind, a cop in a suit himself. It was respect, fear, usefulness, comfort, family, identity. It was all too much to expect him to stand up against them for emergency abstractions that might never happen.

ANTHONY FUSCO WORKED for Rudy Giuliani for six months. As chief of the department, he and Giuliani conferred at major fires. They talked in March 1994 at a blaze in the village that took the lives of three firefighters, suddenly incinerating them as they stood on the second floor of a three-story walk-up apartment building. The captain of that unit, John Drennan, burned over 50 percent of his body, hung on in the hospital for 40 days before dying, and the mayor and Fusco visited him often, interacting again. They met at a three-alarm fire in Harlem. "As soon as I had the opportunity to get away from the command post," Fusco recalled, "I would brief him. He was a very intense manager. He had his hands in everything. He wanted to be briefed. Most of the interaction was between him and the commissioner, but if there was a press conference and there was a briefing that was fire-related, those were the times I would interact with him."

But they never discussed the '93 bombing nor the hundreds of pages of reforms proposed by Fusco and his colleagues. "I don't think that issue—of what we should do with things like that—ever surfaced in any kind of true discussions," said Fusco years later. It didn't come up with Giuliani, Safir, or anyone at City Hall. No one talked to him about terrorism, OEM, helicopter training, radios, defend-in-place fire strategies, joint command posts, the WTC, fire codes at Port Authority facilities, elevator protocols, the disabled, fifth alarms, heavyweight fire gear, or any of the other meaty issues raised in his after-action reports.

"I don't think anything really changed," says Fusco, focusing particularly on his recommendation that a ranking Police Department representative remain at a fire incident command post, just one measure of the unified command failings of the Giuliani years. "Once I left the department, I'm out of it. You had a lot of initiatives, but whether they carried that out after you left, that's questionable."

There was no better example of City Hall laxity over these years than on the question of helicopter rescues, the controversy that almost led to blows between badges in 1993. Those rescues had so convinced Ray Kelly of the usefulness of this tactic that afterward he sent police helicopters around the city to photograph skyscraper roofs in preparation for possible rescues. But Giuliani issued a protocol in 1994 that said only firefighters could do rescues at a fire. Bratton was commissioner, and the decision provoked no noticeable ire within the new administration. It went a step beyond Fusco's proposed joint exercises, which he pushed as a way of getting fire and police rescuers used to working together. Under Giuliani's order, police pilots were supposed to go to designated Manhattan heliports and wait for firefighters to meet them there. Instead of the skilled Emergency Service Unit that rappelled onto the North Tower by rope in 1993 and chopped down the antennas obstructing the helipad, firefighters were to be placed on the roofs.

Greg Semendinger, a police pilot who did rescues in '93 and returned on 9/11, called the protocols "silly." There was no reason, he maintained, "why a helicopter that can take people off a roof would be sent to the opposite side of the city just to get firemen. When you're in a helicopter, time is of the essence, and we should respond to the scene immediately, and try to do whatever we can." There were supposed to be joint training runs, as Fusco had urged, but all they did, says Semendinger, was "a lot of orientation flights." Pilots flew to the West Side, picked up a Manhattan fire company, and took them for a ride. There were no landings and drills. Firefighters weren't taught how to rappel onto roofs, as emergency service cops had done in 1993. Semendinger, whose father was a firefighter, said the FDNY was "always angry" that they didn't have their own copters and "they didn't like the idea that they had to call us to do a rescue." That's why "they didn't request" helicopter help in '93, says Semendinger, "and didn't again in 2001."

And Semendinger believes helicopters could have saved a few dozen people, using 250-foot hoists with folding seats, on 9/11. Hovering over the towers, Semendinger said the northwest corner of the North Tower was virtually smoke-free, and that rotating helicopter hoists could've gotten "a lot of people off" the roof. At another point, a roof rappel team was assembled and a helicopter was dispatched to pick them up. The main reason no rescues were attempted, however, was because "there was nobody on the roof," Semendinger said. And the roof was empty because of a triple system of locked doors installed since 1993. One door could be opened only from a 22nd-floor Security Command Center, an "improvement" the Port Authority attributed to the '93 bombing, though it actually made the rescues that occurred then impossible. The doors automatically unlocked when the power went off in 1993, but even when a decision was made to release them on 9/11, the software malfunctioned and the doors remained shut. The city building code required roof access and automatic unlocks in emergencies, but the authority was exempt from the code, and the Fire Department acquiesced to, if not urged, the roof shutdown.

The Port Authority's Alan Reiss says that helicopter rescues "were considered and dismissed" during a review by the police and fire departments after 1993, and once the city decided they weren't "practical," the authority moved ahead with its lockout plan. Fire Commissioner Tom Von Essen offered a similar history: "The fire department and the trade center were on board in not using the roof for a rescue. That was an agreed-upon policy. This is not something they wanted to do. And the doors were locked for a reason." Asked if this "hard, formal policy" was a consequence of the 1993 rescues, Von Essen said, "It might well have been. The chiefs, in a sit-down with the Port Authority, come up with something; I would have just left it." Neither the FDNY's high-rise fire plan nor the Port Authority's evacuation plan contained a single sentence about roof rescues. Semendinger also says the roof was even more cluttered with antennas in 2001 than it was in 1993, and that they also hampered landings. The Port Authority "was told" he says, to get them out of the way, but the Fire Department's agreement to block the rescues contributed to the authority's indifference about the antennas.

One top police commander barred landings on 9/11, and another ruled out rappeling shortly before the first tower collapsed, but hoists remained an option. The 9/11 Commission's John Farmer concluded, "I'm sure the pilots want to believe that hoists could be done, but I doubt it." The National Institute of Standards and Technology didn't eliminate the possibility that hoists could have worked, but observed that "a very small fraction of the large number of people trapped above the impact zone could have been rescued." The NYPD says that whether rescues could have been achieved is a "moot question," refusing to take a position.

Buzzer locks that no longer automatically unlocked, Giuliani protocols shaped by turf tensions, Port Authority and Fire Department collusion, and profitable rooftop antenna rentals made rescues impossible before the first plane hit. The legitimate doubts about the feasibility of 9/11 rescues miss the point. Though Los Angeles has made helicopters a cornerstone of their emergency response, and though they helped here in 1993, NIST says they weren't "intended to be an option" here, and that was a fixed policy choice long before 9/11. It was settled as soon as Rudy Giuliani took office and never reopened. Jerry Hauer, with his vague and unenforceable interagency powers, wouldn't touch that hot potato. And since there was never an interagency high-rise fire drill or plan, no one ever tested the protocol Giuliani released seven years before 9/11.

As upset as Chief Fusco was in 1993, he tried to keep rooftop rescues, with training and joint participation, in the plans. In fact, the Fire Department circular Fusco helped draft said that helicopters might be "the only access available" in some fire circumstances, a valuable way to rescue "large numbers of people on the roof." But, like so many of the other urgent lessons Fusco raised in 1993, a working roof rescue strategy never took shape.

FUSCO WAS ON a golf course at Myrtle Beach in South Carolina when the first plane hit on 9/11. Eleven other retired chiefs were with him at an annual outing. Almost all of them had children or other relatives in the department; Fusco had two nephews. The group went to the clubhouse and watched reruns of the attack and collapses and made a decision: "We gotta get back."

Planes were grounded, so the chiefs came back in three rented minivans, four per van. By the time Fusco got home, his wife had a list of the 260 firefighters that Fusco knew who were missing. Fusco rushed down to fire headquarters the next day and tried to help set up "places where people could get information." A fire captain drove him to the site. "I was not prepared for what it looked like," Fusco said. "I was standing in the same spot where I stood when I commanded the '93 job. And the huge buildings, and the hotel were just . . . I just couldn't believe it. I was very depressed for a long time. I wound up with a hiatal hernia and I didn't even realize it was from the aggravation. My doctor knew I had been the chief. He said, 'Don't you know what it's from?' I said no. 'You're burning acid every day because of the incident.' "

What was also eating away at him were the missed opportunities. "It's everybody's job to implement these changes," he says now. "The recommendations come and people determine whether it's truly necessary or not. If you feel it's necessary, you still might not always go through with it. I know they ran into some of the same problems with the second incident that I had. The second incident proves again that everything that went wrong in the first incident went wrong in the second incident. I don't think it got addressed properly."

The National Institute of Standards and Technology published an unnoticed appendix listing nine recommendations from '93 and footnoting them simply as Fusco 1 through 9. In muted, sometimes obscure language, it did not identify a single recommendation that resulted in improvement, often stating flatly that the proposed change was not made.

CHAPTER 6

RUDY'S BUNKER

"CITY READYING GIULIANI's high-tech B'klyn bunker" announced the *New York Post* on March 13, 1995. It was four years before the mayor's new, state-of-the-art command center would actually be built, and when it happened it would not be in Brooklyn after all. But from then on, every tabloid story would brand the planned new command center "Rudy's bunker." The name "Giuliani" and the word "bunker" just seemed to go together. And the story, as it played out, would almost be operatic: dubbed a "safe haven," the bunker was ultimately built inside a notoriously targeted complex. Extended into the clouds, it defied the underground habitat of everyone's expectation. A supposed arsenal of celebrated preparedness, it had to be evacuated at the moment of gravest danger to the city. A stunning technological achievement, its misplaced fuel system may have brought down the 47-story tower it called home. And in a final epic twist of destiny, the bunker's demise triggered the redemption of its inventor.

The timing of the *Post* headline was striking, since Giuliani had yet to say anything publicly about the possibility of the new mayor's Office of Emergency Management, which would, of course, wind up building and running the bunker. The interviews for an OEM director had just begun, without any press notice. Yet even at this early stage, top city officials were willing to talk to the *Post* about the concept and possible location of a command center.

The *Post's* prime source of information was Ralph Balzano, the commissioner of the Department of Information Technology and Telecommunications. His agency was about to move into a new five-story, $55 million technology center in Brooklyn—built to order for the city by a private developer. Balzano told the *Post* that, in addition to his agency, the city planned to put a "nerve center, control center, very much like NORAD" in the tech center, which was part of a sprawling complex of office towers near the Manhattan Bridge called Metrotech. The first deputy mayor, Peter Powers, was also quoted. "This is an advanced technology building that has incredible security," he said.

Balzano actually used the B word. "The mayor and his staff can operate from a bunker-like situation," he said. "God forbid, our power plant gets hit, or there's some kind of catastrophe like the twin towers, the mayor will be able to direct out of that building public health, public safety, all EMS services, sanitation." The *Post* said the building, with backup generators for two days and its own water supply, was "designed to withstand fire, flood, and a terrorist bombing on the scale of the World Trade Center disaster." Balzano told the paper that Giuliani would be able to videoconference with Washington, address the city on live TV feeds, and rely on duplicate phone lines and compartmentalized power in the facility, which also housed the Police Department's 9-1-1 system.

Although it hadn't been announced yet, the Fire Department and Emergency Medical Services had also decided to build their own new headquarters at Metrotech—another reason the site was seen by some as perfect for emergency coordination. The new $62 million, eight-story Fire Department headquarters would become home to 1,100 employees by 1997. The plan was for the city's computer and communications network, 9-1-1, emergency medical system, Fire Department, and command center to occupy a protected enclave together.

As suddenly as the idea hit the press, it vanished. Yet it remained in planning discussions. On February 14, 1996, when the city had finally found a director for the mayor's new Office of Emergency Management, and Jerry Hauer was in his third week on the job, he wrote a detailed memo to Powers entitled "Location of Emergency Management Command Center." The only site Hauer examined was Metrotech, which he decided "could meet our needs."

The advantages were striking—particularly when compared with where the command center actually wound up, at 7 WTC. Hauer said the Metrotech center "could be available within six months." (It took three plus years to build at 7 WTC.) The renovation funding was left over in the budget for Metrotech, and no rent would be charged, since it was space the city had already leased. (The bill for 7 WTC's rent and renovations tallied over $61 million.) Hauer said the Metrotech building was secure and, prophetically, "not as visible a target as buildings in Lower Manhattan." He also said the building had "all the backup and redundancy necessary in the event of utility failure," a requirement never fully achieved at 7 WTC despite expensive and dangerous efforts to do it. With 9-1-1 and fire/EMS communications scheduled to move there, he concluded that it would serve "as a focal point for public safety activity."

Hauer also presented the cons, and most of them involved the mayor himself. The first one was that "traveling to Metrotech will consume more of the Mayor's time." While Giuliani could fly there by helicopter or take a Police Department harbor unit, Hauer noted that, during a blizzard or hurricane, "flying would be difficult" or "wave heights" might prevent a boat launch. "Subway access would be possible, but the mayor would have to be willing to do this," Hauer observed. "Ground transport" to the new facility, which was at the foot of the bridge, was also feasible, said Hauer. In the most extraordinary circumstances, the mayor could communicate with the facility by videoconferencing from City Hall. Last, Hauer said the building was vulnerable to a truck bomb, but that police cars and strategic street barriers could protect it. "I believe the cons are not insurmountable," the memo concluded. "The real issue is whether or not the mayor wants to go across the river to manage an incident. If he is willing to do this, Metrotech is a good alternative."

Hauer continued to raise the question of a Metrotech siting—at an August 16, 1996, meeting with Powers and an August 26 meeting with Giuliani. Balzano did the same: "Metrotech was a hardened site," he said years later. "It was created to be a disaster center. We had a top of the line command center. I had conversations with Peter about putting the center there. Peter was supportive. There would be no disruption in service during disasters." But the conversations ended in '96. Recalling his conversations with Giuliani counsel Denny Young, Hauer said he discovered the problem: "Denny said it had to be within walking distance of City Hall. I talked about an underground facility in Van Courtland Park in the Bronx. I talked about available facilities in midtown. I talked about Governor's Island. We could do something at Metrotech or in Queens, I said. Denny made it very clear to me that the mayor didn't want to get on a boat. He didn't want the mayor walking across the bridge. The mayor wants to be able to walk to this facility quickly. That's it. I said to the mayor, based on your requirement of walking distance, here's the area we can do it, all within a narrow range of City Hall. He said fine."

Within those parameters, Hauer began working with the city's Department of Citywide Administrative Services, which handled all contracting and leasing for city agencies.[1] Hauer and agency representatives looked at city-owned space in the Municipal Building and a building right next to police headquarters (he'd already visited many other sites on his own). But the decision to put it near City Hall created a new problem: most of Lower Manhattan is under the floodplain, with hundred-year-old water mains. That's why the Police Department had rejected Lower Manhattan as a site for what it once hoped would be its own new command center, insisting that it had to be "out of flood-prone areas." Hauer didn't have that choice. Since the mayor had decided that the center had to be near City Hall, Hauer was convinced it couldn't be underground—for fear of flooding the facility or its backup power. Given the two new priorities of walking distance and floodplain standards, the search was inevitably being pushed toward the sky.

One other factor framed the City Hall hunt for a site, especially from the Department of Citywide Administrative Services' point of view. Everyone understood that this was a special project. DCAS routinely found and renovated space for city agencies, usually servicing an obscure assistant commissioner from the Department of Human Resources or Health whose job it was to count cubicles and obsess about lighting. One administrator who oversaw this project, however, testified in a subsequent civil suit involving the collapse of 7 WTC that the command center was a far different case. Asked if anything distinguished this rental from "other situations over the years," Glenn Pymento said, "The only difference, if there was a difference, is that there were very senior city officials that were involved in this project." Pressed as to whether this meant the mayor in particular, Pymento fumbled, "Among others, among others." It was, he said, "called high priority, high profile."[2]

DCAS was determined to do it up big, and no one was more determined than Commissioner Bill Diamond. The millionaire son of an East Side real estate titan, Diamond was a financier and player in the Manhattan Republican Party. He had spent much of his life in patronage positions. For virtually the entire 12 years of the Reagan and George H.W. Bush administrations, Diamond had held the politically pivotal position of regional administrator for the General Services Administration. That put him in charge of contracting and leasing in the Northeast, operating out of New York City. His agency's award of a $241 million construction contract for a new federal courthouse in Manhattan to a partnership of Bechtel and Park Tower Development, both major Republican bastions, got the attention of newspapers and investigators. So did multi-million-dollar dealings with a Pennsylvania-based company called Linpro—especially a $276 million contract for a federal office building that was tied to the courthouse project. Shortly after Giuliani named Diamond to head DCAS, *Newsday* reported that he'd destroyed his records the day he left his old job. "They were just ripped to pieces and thrown out," said Barbara Gerwin, the regional General Services counsel. U.S. Senator Howard Metzenbaum wrote the Justice Department calling for a probe and specifically citing Diamond's "destruction of official records," but the controversy soon came to an unexplained end.

In 1989, Diamond started contributing to Giuliani's campaign committees, hitting contribution limits more than once. He and his two wives combined over the years, continuing right up to 2005, to give $36,000 to every political committee Giuliani has ever formed. Marylin Diamond, his first wife, was elected to the Civil Court from Manhattan's East Side in 1990 on the Republican, Conservative, and Liberal Party lines. The couple had already split, apparently amicably, because Bill was at her swearing-in. Diamond himself had run for public office in the 1970s, but became just one more frustrated Manhattan Republican. Giuliani presided at Gracie Mansion for Diamond's second wedding in November 1996—to former fashion designer Regine Traulsen, the daughter of a Morocco food exporter, who traveled between her homes in New York, Paris, and Southhampton and was described in press clips as a Republican fund-raiser. In the summer of 1995, the *New York Times* described a $1,000-a-plate dinner Diamond and Traulsen threw in their Hamptons home for Governor Pataki, noting that the couple then left for a nearby party for the mayor.

The broker Diamond's agency assigned to find space for the command center was at a national shop with a tiny New York presence, CB Real Estate Group. On its board of directors were three Republican giants, including the former counsel to the Reagan/Bush campaign, Stanton Anderson, who'd gotten to know Giuliani when he worked for the Reagan administration in Washington in the early '80s. Anderson's law firm also did work for the company. Fred Malek, the campaign manager for Bush's reelection campaign in 1992, was also a longtime director. James Didion, CEO of the company at the time, says now, "I am aware of relationships that occurred with city agencies and there was a relationship with City Hall. But I don't know what kind, and I don't know who was involved."

Hauer remembers that it was "a young, blond-headed kid" from CB Real Estate Group that first brought him to look at 7 WTC in 1996. He also recalls that Diamond himself was quite enthusiastic about 7 WTC, which splendidly fit the bill of a high-profile building for a high-profile project. "Diamond said he knew the building," said Hauer, who toured it with Diamond. "He did everything he could to facilitate it."

Diamond sure did know the building. In a bizarre turn of fate, Giuliani had once emptied the building and Bill Diamond had filled it up again.

Like everything else at the trade center complex, the Port Authority owned the land underneath 7 WTC. But in the early '80s, the authority had given developer Larry Silverstein a 99-year lease to build a tower on it. Silverstein's prize tenant was slated to be Drexel Burnham, the investment firm that was one of the giants of the Wall Street. But when Rudy Giuliani was Manhattan U.S. attorney in the 1980s, he had turned his prosecutorial sights on Drexel, and by the time Giuliani was done, the indicted company was on the ropes. Dying with it was its agreement to lease a new headquarters at 7 WTC. Silverstein's biggest project began to look like his biggest blunder. He did get another investment house to take some of the space before the building opened in 1988, but gobs of the rest went begging. But happily for Silverstein, during Diamond's tenure at the General Services Administration, nine federal agencies took space in the building, agreeing to pay rents that were often an overpriced embarrassment. Cornelius Lynch, who handled rentals for Silverstein at 7 WTC for six years, recalls, "GSA pushed prospective tenants to us. They liked 7 WTC."

The lure of 7 WTC for Diamond was never clear, but the General Accounting Office, Internal Revenue Service, Equal Employment Opportunity Commission, Securities and Exchange Commission, Federal Home Loan Bank, Defense Department, Treasury Department, CIA, and Secret Service became tenants, some consuming more than two floors. Despite a dramatic downturn in the real estate market, Diamond's agency was paying $56 and $59 a square foot at the same time that city officials were quoted in news stories saying that they were getting nearby office space for $17.50. But as much Silverstein space as GSA was consuming, no one ever rented the 23rd floor of the building. Erected with high ceilings as a trading floor for Drexel, it was vacant for a decade. Instead of Giuliani-target Michael Milken reigning supreme on Drexel's trading floor, Giuliani himself would reign as the commander-in-chief of a state-of-the-art command center.

Hauer liked the space and went to the mayor. "We were very nervous about it at first," Hauer testified in the suit, "because the property was Port Authority property, and I knew how Rudy felt about everything and anything having to do with the Authority. There was a big battle going on." Hauer was also uncertain how the mayor felt about Silverstein, a power broker on the city's real estate and political scene. "When I brought up Silverstein's name, Rudy called him a son-of-a-bitch," said Hauer. "I asked are we okay in doing this? He said go ahead. Work with Diamond."

Silverstein had been a substantial supporter of Ed Koch in the 1989 mayoral race, even making a $25,000 loan to him that had to be returned because it violated city finance rules. In 1993, Giuliani went to Silverstein's apartment for breakfast, seeking his support, recalled Silverstein executive Cornelius Lynch. But, he said, Silverstein, whose daughter is gay, was "offended" by Giuliani's positions on gay issues, which had taken a turn to the right, especially over gays marching in the Saint Patrick's Day parade.[3] Even after the breakfast courtship, Silverstein stayed out of the election, giving nary a dime to either candidate. Giuliani's son-of-a-bitch tag might well have been prompted by the lingering memory of that breakfast.

The relationship, however, was undergoing a sea change. In October 1994, Silverstein began donating to the Giuliani committee, at first just a meager $500. Three months later, he kicked in a thousand. By November 1996, he and his wife, Klara, and Silverstein Properties had contributed $17,500—$4,800 more than the legal limit (when a newspaper noticed it, the excess was returned). To make his allegiance crystal clear, Silverstein also stopped giving to Ruth Messinger, the Manhattan borough president to whom he'd contributed regularly for years. The Democratic nominee against Giuliani in 1997, Messinger was stunned. "We both have gay daughters and we'd talked about that over the years," she said. "We got along well." Suddenly, even though most of his holdings were in her borough, she couldn't get him on the phone. The Giuliani camp was notorious for punishing those who gave to his opponents.

Silverstein hosted a $1,000-a-couple fund-raiser for Giuliani aboard his 130-foot yacht, the *Silver Shalis*, docked at the Chelsea Piers in June 1996, collecting another $36,100 from his invited friends. When Giuliani decided in 1997 to field a Republican candidate against incumbent Public Advocate Mark Green, who was a prohibitive favorite to win reelection, Silverstein kicked $1,000 into the kitty of Giuliani's designated loser, a sure sign of just how wired into the Giuliani camp he was. When Giuliani, unable to run for reelection and uncertain of his future, formed an exploratory federal committee in 1998, Silverstein and an executive of his company gave another $3,000.

Just as suddenly, the developer started giving to Bill Diamond's lifelong favorite charity, the Manhattan Republican Party, donating $1,000 in February 1998, just as the city and Silverstein were working out the final terms of the command center lease signed in March. (Diamond has acknowledged knowing Silverstein, but minimizes the relationship.) By then, though, Diamond's principal vehicle for political upward mobility was no longer the local Republican Party; it was Giuliani himself. Diamond's daughter got a job in the administration as counsel to the Department of Cultural Affairs. Maury Satin, who was Diamond's chief of staff and wound up marrying his daughter, was named to run the Off-Track Betting Corporation, a prime patronage plum. He later joined Giuliani's consulting firm.

Silverstein's contribution to Giuliani's exploratory committee in 1998 was made at a party at Howard Rubenstein's 5th Avenue home. Silverstein had long been one of the power-broker public relations czar's top clients, and Rubenstein had begun talking publicly as early as 1996 about the relationship between the developer and the mayor, confirming a private breakfast Silverstein, Giuliani, and two others had had at Gracie Mansion. In addition to Silverstein, two top executives of the construction company he was using to build the command center, Jay Koven and Jack Shafran, gave $4,000 at the party, bringing the Silverstein-connected total to at least $7,000. Shafran remembered going. "Attendance was obligatory," he said. "The invitation meant we were expected to give a contribution. I remember everybody having their picture taken with Giuliani." The timing couldn't have been queasier. Two weeks before the event, Silverstein had signed the lease for the command center. The day before, the budget office had rushed through approvals for $12.6 million, the initial renovation cost. The budget office's okay—granted, even though Silverstein had yet to supply the legally required scope of work—allowed Koven's company to register its contract with the city.

Shortly after the party, Rubenstein began sharing office space with Peter Powers, the Giuliani First Friend who'd stepped down as deputy mayor in 1996. Powers soon launched his own "strategic consulting business," frequently advising clients such as BAA, who did business with City Hall, and Rubenstein recalls that "Silverstein was one of Peter's first clients." Rubenstein acknowledges referring clients to Powers, but can't recall if Silverstein was one. A law firm with close ties to the mayor—Proskauer Rose—was also representing Silverstein, says John Gross, a partner there, who personally began working with the developer on insurance issues after 9/11. Gross, who has been the treasurer of every Giuliani campaign committee and has known him since the 1970s, declined to say how long his firm had represented Silverstein or on what matters.

In July 1999, just a month after the command center was opened, Silverstein hosted another boat fund-raiser for Giuliani, raising $100,000 for his prospective Senate race against Hillary Clinton. Jay Koven, whose firm had just finished the command center, contributed another $1,000, as his father had at the 1996 *Silver Shalis* event. The 1999 gala remained a secret until late August, when reporters learned that Silverstein was hosting a similar event for Clinton and wrote about the dueling dinners. One reason for the secrecy was that WNBC reporter Ti-Hua Chang had aired a two-part series in June, raising questions about the possible link between Silverstein's campaign support for Giuliani and the command center deal. Giuliani told Chang on camera that Silverstein "donates money to everyone"—something Ruth Messinger would have been surprised to hear. The mayor also said, "I had no knowledge that the bidding process even included him." Chang reported that there *was* no bidding process. But Chang did not know that Hauer had told Giuliani about Silverstein's involvement before any negotiations for the lease even began, belying the Giuliani denial. The administration was so upset by Chang's reporting that it barred him from the media opening of the command center. The mayor's friends certainly wouldn't have wanted Chang and his camera crew to show up at Silverstein's second Chelsea Piers fund-raiser.

When the city honed in on Silverstein's long-vacant floor at 7 WTC, he was a developer struggling through hard times. He had not put a shovel in the ground since he started building 7 WTC in the '80s. His lenders took over his Times Square Embassy Suites development and his office building at 120 Broadway in the early '90s. He was forced out of a mall he was redeveloping on West 33rd Street. An apartment complex of his at 42nd Street had been stalled for a decade. One of the reasons Silverstein did not donate to anyone in the 1989 and 1993 general elections may well have been that he didn't expect to do much business with the winner, whoever it was. By 1996, however, he was making a comeback, and a relationship with the man likely to be mayor for five more years was sound business strategy. The command center and the contributions launched that relationship and, by 1999, Silverstein's 42nd Street project was getting city subsidies.

Meanwhile, the invisible combination of Bill Diamond's history with 7 WTC, and Larry Silverstein's intricate new relationship with the Giuliani administration, had resulted in a decision to locate the city's command center high above the one spot on American soil that had been the target of terrorist attack. And when the city had to respond to its deadliest threat, the reason there was no operational center was rooted in Silverstein's seduction, Diamond's predilections, dubious walking-distance and floodplain requirements, and high-profile expectations.

WHEN HAUER WROTE his Metrotech memo in 1996, he estimated that all the Office of Emergency Management needed was 15,000 to 17,000 square feet for a command center, including a private office for the mayor, a pressroom, 10 other small offices, and room for up to 40 agencies in the center itself. A year later, when the first meetings about the renovation of 7 WTC began, the city's plan called for 46,000 square feet, eventually hitting over 50,000. That's what they rented because that's what Larry Silverstein had to rent—the size of the 23rd floor, with 4,000 square feet added for the backup power system and roof antennas. A tripling of OEM's square footage was just one measure of how the scale and scope of the project expanded to fit the space rather than the needs determining the site.

Under normal circumstances, when an agency wants to rent space, its representatives will work with the Department of Citywide Administrative Services to produce what's called a "program"—which describes in detail the desired features of the prospective space. They then go out and look for a rental that meets the program requirements. Things worked a little differently with the command center, Hauer said in his court testimony, "Once we found the space, they developed the document."

The program was actually developed by an outside firm called Jefferson Inc., which was selected either by Silverstein or by CB Real Estate Group, according to city officials. Richard Ramos, DCAS's executive director of space design, was asked if any program, Jefferson's or the city's, was used in the site selection for the command center. He replied, "I don't believe so." Asked if he, the initial project director, was involved in the original site selection, Ramos said, "I don't believe so," adding that they had "fast-forwarded to WTC 7."[4] It was as if the city picked the winner, then drafted the bid. In fact, the bid, or program, was an after-the-fact concoction of a consultant picked by the winner, Silverstein, or the broker, CB, whose fees were paid by the winner.

What was even stranger about this program was that it was dated July 15, 1997, though 7 WTC had been selected as the site several months before that. Instead of launching a search for space, as a program is supposed to do, this one was dated the same day that Silverstein actually signed and filed a lease with the city. That made the release of the "bid" and the submission of the contract absolutely simultaneous—about as unusual a confluence of events as had ever happened in the coincidence-packed history of city contracting. In fact, on July 14, CB Real Estate Group submitted drafts of all the legal letters and resolutions necessary for the city to approve the lease. The documents had the City of New York letterhead on the top and CB's letterhead on the bottom. That juxtaposition of events made it clear that the Jefferson program was merely one more document prepared at Silverstein's behest to meet a requirement for city approval of the lease. Indeed, DCAS said in a note to the budget office that accompanied the Jefferson analysis that it was developed "to expedite this fast-track project."

The collection of submissions began a review process that would include budget, city planning, and neighborhood approvals, and take until an anticipated October 15 for approval, but the rushed project immediately hit a snag. The budget office questioned the lease because it was unaccompanied by any of the required construction estimates. City Hall, in the throes of Giuliani's reelection, pulled back, perhaps wary of igniting the media firestorm that the whole deal eventually provoked.

The administration waited until November 5, the day after Giuliani's reelection, to start the formal approval process again. Giuliani himself was taking a victory lap around the city, coatless in an open-air bus, starting his morning on *Today* and finishing on *Letterman*. That morning, DCAS quietly sent the City Planning Commission a notice of intent to acquire office space for the Office of Emergency Management. The notice conceded that the submission was not "indicated in a Citywide Statement of Needs," as is routine, because the requirements for the facility "were only recently determined," an assertion contradicted by Jerry Hauer's nearly two-year search. It did accurately reveal that "the proposed site was selected due to its proximity to City Hall."

The next day, for the first time, Giuliani allowed the press to attend his early morning staff meeting, with 16 aides gathered around a table discussing such items as the takeover of the airports from the Port Authority. "We'll be more open," the reelected mayor smiled. "This is our first day of real business again. I thought it would be interesting for you to see it and get a sense of how government operates." But later the same day, when Giuliani met with the Office of Emergency Management to examine the tentative plans for the command center, mum was still very much the word. Memos prepared by Swanke Hayden, the architect drafting plans for Silverstein, indicate that they had delivered two "scenarios" for the facility to the Office of Emergency Management at 9 A.M. and that OEM submitted the alternatives to Giuliani in the afternoon.

The scheme that came out of the discussions that day focused in part on the mayor's needs. He was to get an 1,100-square-foot suite with a pullout sofa bed, private pantry, private toilet and shower, and conference/lounge-type room for his security and assistants. Access to his suite was to be sealed. His conference room required a projection screen, chart rails, marker board, and pinup-type wall panels. The minutes of the meeting also revealed a major quibble with the layout of the pressroom. The eight-inch-high platform in the back of the press area, allowing better views for the cameras, drew expressions of "concern that the Mayor would be standing at an elevation lower than the press" and that reporters would be looking down at Giuliani.

Officials on the local community board and the borough president's office had no objection to the command center—possibly because the notice sent to them simply described it as "office space." The planning commission approved the lease on December 3, after a pro forma public hearing. The whole project was slipping in under the radar. On March 25, the lease was formally executed. The escalating rent over 20 years, including the roof, tallied $37 million. The initial renovation work cost $14.2 million, with Silverstein contributing $1.6 million. But $2.5 million in payments to Silverstein for "excess work," and millions more in technological upgrades, pushed the hard costs to $61.5 million by 2001—not the $13 million Giuliani always claimed. With staff and operational costs, the total rose to $70 million for the first five years.

Still operating in secrecy, the budget office inserted in an April 1998 report on the executive budget for the next fiscal year language that simply said: "Mayor's Office of Emergency Management. Design $1.5 million. Construction $15,150,000." It did not describe the project, and since no City Council committee had yet taken responsibility for overseeing the new two-person OEM, the entry passed unnoticed in the council. By putting it in an April budget summary, the item became an "existing project" when the council focused on the budget for the next fiscal year, which started July 1. Council memos indicated that it only "looked at new funding," not "supposedly existing projects," and that it consequently missed the hard-to-find expenditure. "The manner in which the project was added," a belated council memo concluded, "clearly indicates an intent to avoid scrutiny." It was buried on page 995 of the June capital budget, as classified as the blind-ad search for OEM director had been years earlier.

One enemy of the project found out all about it, though: Howard Safir. A high-ranking city official who sat in on the command center discussions with the mayor recalled, "Safir was vehement. He called it a horrible place. He and Richie Sheirer even called it Ground Zero. They invoked the '93 bombing. I remember the meeting vividly."[5] Hauer said, "Safir's biggest argument at the meetings was that the terrorists might shoot a handheld missile at the building."

In June 1998, with the budget under consideration and actual construction getting ready to start at 7 WTC, the *New York Times* finally noticed the command center. A June 13 story by Kit Roane sparked an instant controversy, most of it ridiculing the center as a bunker. "Having tamed squeegee men and cabbies, murderers and muggers," wrote Roane, Giuliani was "now bracing for a whole other order of urban treachery" by building an emergency center "bulletproofed, hardened to withstand bombs and hurricanes, and equipped with food and beds for at least 30 members of his inner circle." Hauer later testified that Safir "planted that story." Roane was a young reporter on the police desk, and that fed City Hall suspicions. But the *Times* had been such a Safir critic, it would have been an unlikely beneficiary of a tip from him. Sources said years later that Roane got it from cops, and confirmed it with NYPD brass.

It's hard to tell how much of Safir's resistance had to do with reasoned concern about the location and how much was simply a continuation of the turf war with OEM. But he was hardly alone. Chief of Department Lou Anemone, too, was a fierce opponent. "Walking distance?" he asked years later. "Why? I've never seen in my life walking distance as some kind of a standard for crisis management. I did a couple of memos against that site, citing the closeness to an intended target, the 23rd floor dangers and hazards. It was a joke." Anemone actually recommended that the command center be located in another borough—like Brooklyn's Metrotech—to make sure government would continue to function should Manhattan be convulsed by an attack. "You don't want to confuse Giuliani with the facts and his 'yes men' would agree with him," Anemone recalled. "In terms of targets, the World Trade Center was number one. I guess you had to be there in 1993 to know how strongly we felt it was the wrong place." Even John O'Neill, the FBI's top counterterrorist chief, quietly opposed the siting, according to his biographer, Murray Weiss. Council Speaker Peter Vallone, ordinarily a close working partner of the mayor's, said at the time, "Look, we have no problems with anything that makes the emergency preparedness system in the city better. We just have a problem with the location."

The Giuliani defense for picking the site despite the obvious security concerns was that the Secret Service was in the building (after 9/11, it was revealed that the CIA was there, too). But the Secret Service and CIA had *offices* there, not command centers charged with managing the city's response to a deadly attack. And they were in the complex before the '93 bombing; they didn't choose it afterward. Indeed, it was ironic that the federal propensity to consume floors of Silverstein space while Bill Diamond was running the General Services Administration became the rationale for the city to take more when Diamond was running DCAS.

The press coverage was devastating. News columnist Michael Daley compared Rudy's bunker with Saddam's, whose $65 million, 11-room, German-built underground facility rested on shock-absorbing springs encased in rubber. One radio show spent four hours lampooning the project and running a name-that-shelter contest, with winners like "The Nut Shell." Even the staid Vallone came up with a zinger: "If he wants to build a bunker for only the people he trusts, all he needs is a phone booth." Elizabeth Kolbert, a *Times* columnist, explained why the story had hit such a chord. "The trouble is that the project seems to speak so clearly to the mayor's fears, and, worse still, his hopes. He has always governed as if from a bunker. On some deep level—well, maybe not so deep—one senses that nothing would please him more than having any antagonist truly worthy of his energies instead of just a ragtag bunch of disgruntled taxi drivers, street artists and unemployed strippers."

All the derision did nothing to slow the project. Construction started in July. Silverstein's handpicked contractor, Ambassador Construction, whose principals were regular Giuliani campaign donors, were paid by Silverstein, who was reimbursed by the city. That meant Ambassador didn't have to go through the city's regular vetting processes. If the contract had required city approval, the mayor's contract office or the comptroller might have been forced to reject it because of a criminal probe involving the company that began in 1996. (Ambassador was named in *Times* stories in 1998 and 1999 as the subject of an investigation by District Attorney Robert Morgenthau's office. In 1999, a design executive pled guilty and was sentenced to three years in prison for taking $1 million in bribes from Ambassador and another firm, but no one from Ambassador was indicted.) Hauer said he wouldn't have hired Ambassador if he had known anything about their criminal problems. In the civil case, he testified that "we let Ambassador do the construction so that we didn't have to bid it out," ceding it to Silverstein.

Giuliani gave Hauer a January 1, 1999, deadline to finish the project, and Hauer technically met it, moving in on December 28, 1998, though the facility was far from ready. In fact, in one more extraordinary breach of city policy, the Office of Emergency Management's masters-of-the-universe command center had no certificate of occupancy or any other equivalent until April 2001, almost two and a half years after it moved in. Because it was on Port Authority land, what was formally required was a letter of completion from the authority. But the authority refused, citing shortcomings, many related to the fuel system. So DCAS executed a new agreement with Silverstein in March 1999 "deleting the requirement" in the lease that Silverstein deliver the "Port Authority sign-offs."

Had the command center been located anywhere other than exempt property like the Port Authority's, the city could not have legally occupied the space without certification by its own Department of Buildings. But, as the city's attorneys pointed out in a post-9/11 brief, "the Port Authority's ownership of the land is a key fact, because its ownership meant that 7 WTC was not subject to the requirements of the New York City Building Code." Instead of following through on the recommendations after the 1993 bombing to put Port Authority property under city codes—embodied in the bill introduced annually by Senator Roy Goodman—Giuliani took advantage of that loophole. Freed from the yoke of its own building code, however, the city then blew off the objections of the authority whose exemption was their escape hatch. It was full-speed-ahead Giuliani at its worst. The agreement with Silverstein even specified that the city "acknowledged that the Port Authority sign-offs may not occur for approximately six to 12 months." In fact, it took twice as long. All that time, the city remained in a highly sensitive facility without any form of construction certification.

The grand opening press releases in June advertised the center's capabilities: weather monitoring, national radar, slosh overland maps with evacuation routes, hospital power grids, fiber-optic cables for media, and geographic information systems. It boasted of the 11,000-gallon backup potable water supply, hurricane wall that could withstand winds of 160 miles per hour, traffic camera feeds from stoplights, 45-seat pressroom, and 6,000-gallon backup fuel tank. Giuliani's office had a humidor for cigars, city monogrammed towels in the bathroom, and mementos from City Hall, including a fire horn, police hats, and fire hats. He had his own elevator. His suite was bulletproofed but not his conference room. DCAS memos to Silverstein's office raised issues such as: "Shouldn't the mayor's elevator lobby have a better rubber floor? Maybe a pattern?" and "Mayor's office should have the best furniture in this project."

"Rudy came up to the command center regularly on weekends when he was in the area," Hauer recalled. "He was there for the West Nile virus, the heat, the Northern Manhattan blackout as well. He could watch the harbor, pull up the Coast Guard cameras. Sometimes he'd come by just to have a cigar."

The line of critics who have blasted the siting of the center is so long that it includes some of the mayor's closest associates—people like Sheirer whom he took to work with him at Giuliani Partners. Yet Sunny Mindel, once the spokeswoman for the city and now the spokeswoman for Giuliani Partners, suggests it's all hindsight. "At the time, given the type of emergencies that could beset a modern urban center, it seemed absolutely appropriate," she said of the location decision in 2002. "No one could have predicted the events of September 11." In fact, Silverstein's property risk assessment report identified the scenario of an aircraft striking a tower as one of the "maximum foreseeable losses" just months before 9/11. A congressional inquiry after 9/11 cited numerous indicators that such attacks were a possible terrorist tactic, including one specific aircraft threat involving the World Trade Center.

But it's the contention that no one could have predicted a terrorist return to the World Trade Center—regardless of the form it took—that is particularly clueless. U.S. Attorney Mary Jo White says, "I didn't think it made any sense to put the command center at 7 WTC, where it was in the zone of likely attack." Two-time police commissioner and onetime security consultant to the Port Authority Ray Kelly adds, "If Giuliani had any sense of the threat, he would have gotten out of the City Hall area. He put it right next to a target. It was just unwise."

WHEN 7 WTC BECAME the only modern American skyscraper ever to collapse due to a fire, it was a surprise to many to learn that the building was actually a blowtorch. It had 16 emergency generators in it, more than in any but a handful of new telecom/Internet towers in New York, and they sat above a Con Ed substation with 109,000 gallons of oil in it. That's apparently why the building had such a disastrous fate on 9/11. Debris rained down on 7 WTC from the towers that morning, as it did on countless neighbors, and none of the others toppled. The Fire Department let the fire burn, understandably unable to lift a hose, but it did the same thing with all the nearby towers, the rest of which burned out and still stood. While no definitive conclusion has been reached in the court cases or various studies, the prevailing view is that Silverstein's tower became its own inferno, consumed by its five diesel fuel systems that ran frightfully close to the core steel trusses. The size and danger of the system was obviously known to the city when it added its own generators and fuel tanks to an already combustible mix.

Citigroup alone had nine generators on the fifth floor for its computerized trading operation. The system had the capacity to pump fuel at a rate of 75 gallons per minute, though the generators consumed only 9 gallons a minute when all were turned on. A minimum of another 18 gallons of fuel was circulating through its pipes, even if none of the rest of the 48-gallon capacity was in use, at a pressure of 50 pounds per square inch. This fuel distribution system was, at points, within feet of the structural trusses that supported the entire 47-story tower. Con Ed attorneys claimed that Citigroup had "created a fuel distribution apparatus which became, in essence, either a 160 megawatt blowtorch pointed at, or a lake of fire beneath, those structural trusses." That was all aboard the welcome wagon at 7 WTC when Rudy Giuliani, mesmerized by the threat of a hurried walk across the Brooklyn Bridge, shook hands with Larry Silverstein.

Citigroup had two 6,000-gallon fuel tanks. Silverstein himself had 24,000 gallons in backup tanks. The Office of Emergency Management was coming in with its own 6,000-gallon reservoir, and unlike the others, it would actually hang 15 feet above ground, a suspended dare to fire safety. Silverstein, his contractor Ambassador, American Express, and the city also had smaller day tanks on upper floors, plus all the fuel in the pipes, leading to the determination that there was a total of 43,284 gallons in the building that day. Had 7 WTC not been exempt from the city code, this complex system might well have become a magnet for building department inspectors who approve fuel tank installations. But as excessive as this fuel load already was, it was the fuel system for the command center that really set off alarm bells. From the moment the final plans were turned over to the Port Authority in early 1998, its engineers had a problem with the placement of the fuel tank—hanging from a mezzanine area far above the first floor. Installing tanks underground protects them from falling debris, and, as the *Times* later put it in a story about 7 WTC, "impedes leaks or tank fires from spreading throughout the building."

The city did not want an underground tank, so Silverstein's engineers came up with the mezzanine compromise. The Port Authority balked, more concerned about the apparent violation of the city building code than the city was. The code permits only tanks of 275 gallons or less to be installed above the lowest story, and it is so wary about even these small tanks that it bars any more than one per floor. In fact, the code provides that tanks with a capacity greater than 275 gallons have to be "buried" inside a building so that the top of the tank is at least two feet below ground level. All of 7 WTC's other tanks that exceeded 275 gallons were buried and met city standards.

The Giuliani administration was determined to put its tank above the floodplain, even though Citgroup, Silverstein, American Express, and Ambassador all thought their concrete-protected underground tanks were secure. Giuliani and Hauer had made the floodplain threat a crucial issue when they argued that the Office of Emergency Management should be taken out of 1 Police Plaza, where the Police Department's backup fuel was stored in the basement. Now they were actually hoisted on their own petard. So the city continued to push for an elevated 6,000-gallon tank. OEM's press release at its 1999 command center opening bragged about this feature, even though, by any fair reasoning, the city was exceeding its own code by 5,725 gallons.

The issue became so politically sensitive that Hauer and Giuliani turned to a deputy mayor who otherwise had nothing to do with the project—Randy Levine, another former prosecutor who ran the city's economic development programs. Levine was chosen because he was the only Giuliani aide with friendly ties to the New York State Republican Party and the Pataki administration. Levine had stayed clear of the war with the Port Authority and had a talking relationship with its executive director, Bob Boyle. "We wanted to insure action," Hauer recalled. "Rudy thought it would be helpful for Randy to break the logjam. Randy felt he could get Bob to do something."

On April 6, 1998, Levine wrote Boyle an extraordinary letter, carefully describing the proposed installation as "near the first floor," and contending that "conservative fire codes" related to the "structure enclosing the tank" were exceeded. Even though Levine acknowledged that the port's engineers "feared" that the placement of the tank on a mechanical platform used by building personnel "would provide a point of fire ignition," he asked Boyle to intercede. Boyle sent the letter back with a handwritten Post-it: "Randy, took care of this. Should be reflected in meeting they are having on 4/15/98." After the Levine letter, Hauer said, "There was some sense of movement." A Port Authority engineering memo noted that it received a "review request on April 14," and concluded that the planned installation was "acceptable in concept," with conditions that could be resolved at a meeting. The tank wound up on the mezzanine, thanks to pressures from the highest levels of the city to violate its own code.

The issue might have ended there except for the fire chief who ran the battalion that would have to fight a fire at 7 WTC. The department had no enforcement power on Port Authority property, but it did have that memo of understanding, which gave it advisory power. Shortly after the Office of Emergency Management moved in, Battalion Commander William Blaich took his own tour of the command center fuel system and didn't like what he saw. His report, filed on March 14, 1999, was entitled "Dangerous Conditions at 7 WTC." He listed three significant code violations, centered around the fact that the 6,000-gallon tank was "installed on steel beams at a height of 15 feet above the floor," and that "the maximum allowable tank size above the lowest floor is 275 gallons." He also cited several reasons to fear leaks from the tank, and noted that "in the event of a leak, fuel oil would fill the elevator shaft" nearby. Anticipating a fire that "would be a disaster," he said the conditions could "severely endanger" firefighters attempting to contain it. On March 25, he added a fourth violation about the piping.

Blaich's 1999 inspection was actually a follow-up to two earlier visits by his command in October 1998 that identified some of the same problems. Hauer testified that he never saw any of the Fire Department reports. But the Port Authority did see them, and that was one of the reasons it had refused to grant a letter of completion. In fact, the authority sent a seven-item list of objections about the tank to OEM in May, and an angry Hauer forwarded it to Bill Diamond. Hauer's letter referred to "serious fire code violations," blamed them on Silverstein's architect, and said he would be asking city attorneys to sue to recoup fees and prevent the architect from being used by the city again. No such suit was ever filed, and Hauer couldn't recall ever getting a response from Diamond. The standoff with the Port Authority continued for another year and, even when the authority finally did sign a letter in 2001, long after Hauer was gone, no one could explain why.

In Con Ed's lawsuit against the city, it argued that nothing was ever done to address the fire violations Blaich had listed, and the city's response in court was simply that the "safety concerns were considered," not even contending that they were meaningfully addressed. U.S. District Court Judge Alvin Hellerstein grilled city attorneys about the fire violations in a 2005 hearing, asking whether the department was "ultimately content with the placement of the tanks." The city lawyer said, "I believe so, your Honor. But there's no final entry that says we've signed off." Hellerstein observed, "The fire department clearly was concerned about the placement of these tanks. The bottom line, let's assume the department did not very much like the idea of these diesel fuel tanks being situated where they were; and OEM, by delegated authority, says, we hear your advice, but we're not following it."

But the city's disdain for fire safety requirements went much further than Hellerstein ever knew. The National Institute of Standards and Technology found that the city's "original request" was to locate the 6,000-gallon tank on the seventh floor—a plan that was blocked by the Fire Department. If 15 feet above ground was a code violation, seven stories was a code obscenity. So the city put its 275-gallon day tank on the seventh floor. NIST's interim reports on the 7 WTC collapse repeatedly mentioned that floor as a possible trigger for the collapse, noting that fires were visibly recorded there all afternoon and that critical transfer trusses were also located there. The only tank on the seventh floor was the city's. The reports also focused on the fifth floor, where a Silverstein tank and the Citigroup's tankless, pressurized system coexisted with vital trusses and columns. NIST concluded: "This region of the building"—floors 5 and 7—"played a key role in destabilizing the remaining core columns." The presumption is that fire brought the building down, and that the fires undermined the columns and trusses on these floors. So what caused fires on floors 5 and 7 fierce enough to destroy columns and trusses? NIST lists "unusual fuel loads" as a "possible contributing factor."

The findings are consistent with FEMA's preliminary hypothesis, issued way back in 2002. FEMA said that "the structural elements most likely to have initiated the observed collapse are the transfer trusses between floors 5 and 7" and that "the loss of structural integrity was the likely result of the weakening caused by fires" on the same floors. FEMA also reported that up to 12,000 gallons from the underground tanks were "lost" that day and could have "spilled into the debris pile." It could not determine definitely which underground tanks were depleted, nor was there any data on the "post-collapse condition of the OEM 6000-gallon tank," which was fed by an underground tank. "Although the total diesel fuel on the premises contained massive potential energy," FEMA concluded, "the best hypothesis"—meaning the distribution system—"has only a low probability of occurrence."

Irwin Cantor, who was the structural engineer for the original construction and for the command center, told the *New York Times* that "diesel-related failure of transfer trusses was a reasonable explanation for the collapse." The only planning commission member who abstained on the command center vote, Cantor said he knew that none of the tanks were envisioned in the original design of the building.

Judge Hellerstein dismissed the key Con Ed case against the city for reasons having nothing to do with the substance of the collapse issue. (Hellerstein decided in January 2006 that the city was performing a civil defense function when it built the center and thus wasn't liable for its own possible negligence.) But the opening paragraph of Hellerstein's decision cited two reasons for the collapse of 7 WTC—fires "unquenched by water" all day that were "fueled by diesel fuel stored in tanks located in 7 WTC," certainly suggesting that either the city's, Citigroup's, or Silverstein's tanks were a likely culprit. Hellerstein allowed the suit against the Port Authority, Silverstein, and Citigroup to continue.

Before the federal courts took charge of all the 9/11 cases, one state supreme court judge in Manhattan, Michael Stallman, did get to issue a couple of decisions on preliminary issues. "It is not for this Court to inquire into the City's decision to store such substances in this location," he ruled. "However, it cannot be disputed that, whatever the rationale, the City caused and created the condition. The City had exclusive knowledge of the circumstances of its secret storage of a large quantity of flammable fuel in an office building open to the public." Citing "the unique circumstances of this case," Judge Stallman found that "the City stored the fuel and knew that the ignition of it by WTC debris had a role in the destruction of 7 WTC."

A couple of months after Stallman's decision, Giuliani Partners announced it was forming a "strategic partnership" with CB Richard Ellis, the successor of CB Real Estate Group and Richard Ellis, which merged just as CB was collecting its fee on the command center in 1998. The press announcement said that Giuliani Partners would be using "its proven crisis management expertise to advise current and future CB Richard Ellis clients on complex issues related to emergency preparedness," including "location and site assessment" as well as fire safety. Rudy Giuliani was quoted as saying that businesses were now looking "to anticipate problems before they occur" and that they want "to do a comprehensive assessment before they move to a different location."

A year later, CB Richard Ellis became the exclusive leasing agent for Larry Silverstein's sparkling new and vacant 7 World Trade Center.

CHAPTER 7

THE SOUND OF SILENCE

WHEN RUDY GIULIANI testified at the 9/11 Commission in May 2004, members of the audience—some of them relatives of dead firefighters—began yelling "What about the radios?"

The radios—the antique "handie-talkie" devices that had served the firefighters so badly during the 102 minutes when good communication was needed most—were a scandal, and one that haunts Giuliani to this day. The bereaved survivors are not the only ones mystified by how New York's Bravest wound up unable to get simple information that could have saved their lives that day.

"What about Motorola?" the families yelled.

Sally Regenhard, who believes that her firefighter son, Christian, died because he didn't hear a radioed evacuation order, was one of the voices in the theater of The New School, where the hearing took place. Giuliani had delivered the eulogy at her son's funeral on October 26, 2001. "In those days, people thought having him was a good thing," Regenhard explains. "I thought it was a good thing."

Regenhard stood on 13th Street outside the school after Giuliani's testimony. She pointed out Bob Leonard, the public relations adviser for Motorola in New York, who had been a spokesman for the Fire Department when it began the process that led to the Motorola contract she believes contributed to her son's death. Leonard's public relations company, Dan Klores Associates, also represented Giuliani

Partners. It represented Miramax Books when it published Giuliani's *Leadership* and Brillstein-Grey Entertainment, the Los Angeles–based agent for Giuliani that produced an HBO 9/11 special featuring Giuliani. It represented Bernie Kerik when he imploded as the nominee for Homeland Security Secretary. Leonard, who also used to work in Giuliani's City Hall press office, was hanging out in the May sun that day with the current Fire Department spokesman. To Regenhard, Leonard embodied at that moment the incestuous ties between Motorola and the administration whose radio business it disturbingly monopolized.

While in fact Leonard's multiplicity of hats was more coincidence than conspiracy, the details of indifference and collusion engulfing the Motorola deal tore at some firefighter families almost as painfully as the recurring memories of their loss. The record of favoritism and failure became a wound that throbbed every time Giuliani was knighted, lionized in cover stories, cheered at conventions, retained for millions as a security consultant, or boosted as America's Churchill on his way to the White House. It was not as if Giuliani himself had taken a dime from Motorola, not even in campaign contributions. But the layers of cozy relationships, and the chronology of deception and connivance that infected his administration's dealings with the company, were just the sort that prosecutor Giuliani once focused on like a laser beam.

Under Giuliani's watch, the Fire Department had tried to equip its firefighters with portable radios from Motorola that didn't actually exist when the FDNY took the first steps to order them. The radios were so "undeveloped," according to a subsequent investigation, that the company didn't even put them on its sales list until months after the order was placed in 1999.[1] When the biggest potential fire department customer in America asked for samples, Motorola couldn't send them even a preproduction model for three months. It took three years for Motorola and the department to actually put the first of the 3,818 new radios in a firefighter's hands. Then they boomeranged instantly, failing to communicate a half dozen "maydays" from a trapped firefighter.

Unlike the handie-talkies the Fire Department had used since the 1960s, these new digital radios, the XTS 3500R, were actually computers that turned the familiar real-life voices of firefighters into urgency-deaf monotones that had a prerecorded and often indecipherable quality to them. No large fire department in this country or anywhere else had or would ever purchase them. The contract for them appeared so wired that only one other company, General Electric/Ericsson, showed up at the 1997 pre-bid conference, and FDNY officials decided it was ineligible to participate.

GE/Ericsson's protest letter to city officials, including Giuliani himself, went unanswered. Warren Stogner, the salesman for Ericsson, charged in the letter that "only Motorola can meet the specifications" because language in the bid specified their product and required radio features proprietary to the company. No one heard him then, but years later, Alan Hevesi, the independently elected city comptroller, examined the process and concluded that there was "a willful attempt to circumvent the contract system," withholding over $8 million still due Motorola. Tom Von Essen, the commissioner who issued the XTS 3500s, was forced by the unheard "maydays" to pull them out of the firehouses and conceded that there were "problems" with them that had to be corrected. Yet Giuliani attacked the motives of the Democratic comptroller as "political," defended his friend and commissioner, called the radios "state of the art," and insisted there was nothing irregular about the contract.

When Von Essen said he would do an "internal investigation" of his own sudden scandal, Giuliani agreed, rejecting any notion of an outside probe, or one conducted by the city's Department of Investigations (DOI). Even after Hevesi formally referred the matter to DOI, the Giuliani-appointed commissioner, Ed Kuriansky, ignored the referral, questioned no one, and subpoenaed no records. Hevesi's office had made only one other referral in eight years, but Kuriansky rebuffed this rare request from the city's charter-empowered monitor of corrupt contracts. And Von Essen quickly ended his "internal" review without releasing any findings.

Once the 3500s were pulled, all the firefighters had to pick up their Saber VHF radios again, estimated by the department and Motorola to be between 15 and 18 years old. In a war on 9/11 between terrorists armed with box cutters and fire chiefs who had to rely on runners for messages, bin Laden wound up with the low-tech edge. That's why the National Institute of Standards and Technology determined that the missed messages on those ancient-mariner radios that day caused the deaths of firefighters. Having done nothing to seek an independent investigation of the radio deal in the months leading up to 9/11, or even to support Hevesi's referral to the Department of Investigations, Von Essen began to say after it how much he'd welcome a probe. But just as Giuliani refused to launch an after-action examination of the 9/11 emergency response, his administration did nothing to probe the radio debacle, either.

Motorola had maintained a death grip on city purchasing long before Rudy Giuliani became mayor, but no previous administration had as much incentive to ride herd on the radio issue. For one thing, no previous administration had learned how badly the firefighters' radios behaved during a terrorist attack. And as a prosecutor in the 1980s, Giuliani made his name by convicting a contracting cabal that had turned the city's Parking Violations Bureau into a racketeering enterprise. The biggest scandal in modern city history, it toppled the two most powerful Democratic bosses in New York and sank the Koch administration in the 1989 election.

Mayor Koch was never directly implicated, but the epitaph for his administration was a passage that closed one book about it: "In the end, Ed Koch had not been different enough. He became the mayor who didn't want to know. Admiring his own performance, he didn't notice anyone else's. While he had been gazing into the mirror, his city had been for sale." Rudy Giuliani was the hero of that book, called *City for Sale*, but that was before he, too, stopped noticing anyone else's performance.[2]

THE LONG TRAIL to the radio disaster of 9/11 starts on September 28, 1990, when the Fire Department produced a seven-page assessment entitled "Justification for New UHF Radio System." Using 1989 statistics—the year Giuliani lost his first mayoral campaign—the department concluded that the system it would still have 11 years later was already "obsolete" and "totally inadequate."

The document laid out how "severely overloaded" the system was, dividing the 2,700 radios in the field by the six usable channels and concluding that there were 450 radios for every channel, "far in excess of the Federal Communications Commission standard of 100 units per channel." It called the overload "a direct threat to property, public safety and firefighter safety," citing the "countless" times that firefighters at a fire were "unable to find air time." The report also indicated that the Fire Department's five dispatch units were limited essentially to the same base channels, causing "direct interference" between the dispatchers who sent trucks to a fire and the firefighters who were at it.

Under the heading "No Expansion Capability with Present VHF System," the department underlined the need to move to UHF, where up to 20 channels could be made available, at least one of which would permit direct communication with the UHF-using Police Department. The report called for replacing the VHF radios "configured in the early 1960s" and a push for new channels allocated by the FCC. It warned, prophetically, that a failure to act represented a "danger" to civilians and members of the department.

Nearly three years later, in the aftermath of the 1993 bombing, Chief of Department Anthony Fusco reinforced the earlier memo. The major problem encountered in responding to the bombing, he wrote, was "the communications situation." His postincident assessment identified the key communication problems: too many firefighters on too few channels, "not enough channels for operational use," an "inability to contact other agencies" such as the police, and "lost messages." His calls for a tone alert on radios, which would allow critical commands to get through by signaling that all other transmissions should cease, was ignored. But his recommendation that "improved handie-talkies" be evaluated was consistent with the assessment that had been going on since 1990. Fusco was optimistic that the "severely overstressed" system—almost precisely what the 1990 report had said—would be resolved by "several initiatives underway to enhance communications."

One of the initiatives, launched in 1992, was an application at the FCC, filed by 12 public safety agencies, to get 120 new UHF channels. It was the pathway to a greatly expanded citywide system that, for the first time, would have police and fire on the same frequencies. Initiated in the Dinkins administration, the petition was finally granted by the FCC in March 1995 and celebrated at a press conference by Giuliani and Police Commissioner Bill Bratton, who had led the fight for it. Bratton had helped secure an agreement with a coalition of New York television broadcasting companies who agreed to give up their claim to Channel 16, an unutilized band of frequencies assigned by the FCC for broadcasting use.

What convinced the broadcasters and the FCC was the decision by Bratton and the Giuliani administration to transform the original petition—which sought permanent control of the frequencies connected to that TV channel—into one that would last only five years. That way if a broadcaster needed the spectrum up the road they could apply for it, and if the city didn't make use of the spectrum it could lose it. That put a premium on the 12 agencies—including some from surrounding suburban counties—getting their communication acts together.

The Police Department had been the driving force in the effort to obtain the spectrum, because its radios were already UHF and on frequencies adjacent to the band the FCC agreed to provide. That meant the police could "utilize their existing radio equipment, with minor, inexpensive modifications on both the existing frequencies and the new spectrum," said the FCC. More than any other single governmental action, this obscure FCC spectrum grant, seized by the police as a golden opportunity, was the reason its radios performed reasonably well on 9/11. The department spent $50 million in the next few years building an infrastructure for its operations on the Channel 16 radio frequencies, which were in the 482- to 488-megahertz range. In post 9/11 submissions to the FCC, the Police Department said it had "invested in transmitters, antennas, repeaters and approximately 25,000 portable and mobile radios" on Channel 16.

The FCC would rule years later, when it made the initial temporary waiver permanent, that "the terrorist attacks of September 11 underscored" the need for the spectrum, which the commission said had become "an essential part" of the city's telecommunications defenses. The city's post-9/11 petition and FCC finding, however, conceded that as useful as the spectrum had been to the police, the Fire Department had failed for years to take advantage of the UHF bonanza approved way back in 1995.

The Fire Department was barely involved in the FCC fight, and Howard Safir, the fire commissioner, didn't even appear at Giuliani's press conference. There, the mayor said that the new Channel 16 frequencies would, "for the first time, allow Police, Fire and Emergency Medical Services to communicate on the same radio frequency." That advantage, known as "interoperability," would ensure "a coordinated response during major incidents and more effective radio communications," the mayor promised.

But it actually took more than two years for the Fire Department to take its first steps toward using the new UHF spectrum. In 1997, the longtime assistant commissioner for communications, Steven Gregory, said in a memo to the city budget office that the Fire Department was "proposing the complete replacement of all existing VHF radios with UHF."[3] Incredibly, the memo lifted the entire rationale for the new system, virtually word for word, from its 1990 predecessor, accurately reflecting how little had changed. The system was still "obsolete," "inadequate," "dangerous," and "severely compromised." (The 1990 warning that the problems due to "extreme overloading" on this system "directly impact the danger to civilians trapped in fires and danger to firefighters themselves," was modified with the word "may.")

Gregory added reasons for urgency in 1997 that didn't apply in 1990, starting with the new spectrum. The earlier memo merely alluded to the city's need for additional UHF channels. Since it now had the Channel 16 frequencies, the new memo stressed the five-year timeline. "If these frequencies are not used in the near future," Gregory wrote, "we would run the very real risk of losing them and never being able to regain them again. This would all but certainly kill the possibility of interoperability, something which FDNY, NYPD and others have strived for many years [sic]."

The new memo also transformed the muted 1990 interest in fire-police interoperability into a crusade. "The ability to coordinate with other city agencies, especially the NYPD," the 1997 letter argued, would be "a monumental improvement in our overall operations." Pointing out that the city's transit police—newly merged with the NYPD—would soon be on UHF, the Gregory memo anticipated the "immediate" ability of the Fire Department to at last be able to talk to police above and below ground. "The lack of interoperability has always been a longstanding operational problem," he wrote, echoing the promise of the 1995 Giuliani press conference.

The other new imperative by 1997 was the need for the firefighters to talk to Emergency Medical Services. Giuliani's most significant Fire Department initiative was to merge the FDNY with EMS, which had historically been a subsidiary of the city's Health and Hospitals Corporation. The EMS had long been on UHF, and Gregory pointed out that the Fire Department's planned conversion would make it possible for the firefighters and ambulance personnel "to immediately communicate with each other." While Gregory said that would have "a significantly positive impact on our overall operations," he did not say the opposite—namely, that the failure to do so would prolong the bizarre circumstance of two wings of the same emergency agency rushing to the same scenes on incompatible radio bands.

As compelling as the recycled 1990 critique and new urgent issues were in Gregory's 1997 memo, it would take another four years before UHF radios would hit firehouses. A department that prided itself on reduced response time to fires couldn't figure out a way to respond to the shrillest sirens about aged radios that were a "threat," even when it was sounding the alarm itself. Tom Von Essen replaced Howard Safir, the department moved into its supercharged Metrotech headquarters, new dispatch facilities were built all over the city, and yet the radio conversion, even faced with a five-year federal deadline, had the feel of a leisurely midday fire-truck trip to the local deli for sandwiches.

The only rush on the radio schedule was the expedited effort to dump Motorola's competitor, General Electric/Ericsson, a company whose radios had beaten number one Motorola in some recent matchups. On September 11, 1996, GE/Ericsson met with fire officials at the department's technical unit office at 104 Duane Street in downtown Manhattan. GE/Ericsson brought its new-generation portable radio, and a department report written later that day listed eight reasons it "was not compliant with current FDNY requirements." None of them involved the core communications function of the radio. Instead, the department wanted a label identifying the batteries, a heavy-duty leather shoulder strap, and a rear hang-up button on the microphone. Even though its own internal memos indicated it was looking to greatly expand the number of channels—to 16 or 20—the department objected to GE/Ericsson's 16-channel frequency selector, insisting on an 8-channel switch. The three most significant questions related to the "submersible integrity" of the radio—that is, whether it could pass a wet test under six feet of water for half an hour. Warren Stogner, GE/Ericsson's chief salesman in the area, was left speechless.

According to minutes of the meeting, the Fire Department's tech director, Don Stanton, closed it by telling Ericsson that the department had to have a new contract in place by February 1997. He explained that the department "cannot be without a means of procuring portable radios for any extended period of time" and that the current contract expired in January. This urgency disappeared, however, as soon as Ericsson did. Once Ericsson was forced out of the competition a few months later, the FDNY took a year to sign the Motorola contract, and four more years elapsed before the new radios appeared.

But it was not just the brief time period and the strange objections that troubled Stogner. In the mid-'80s he'd met Stanton's father, a respected technician who'd made all the radio purchasing decisions at Emergency Medical Services when it was a wing of the city's Health and Hospitals Corporation. "He never bought anything but Motorola," Stogner recalled. "I could barely get in the door. I was afraid it came with the genes."[4] Stanton told him that if Ericsson did manage to modify its radio to meet these objections, there might be new ones. This was only a "cursory review," Stanton said, and the radios would still have to undergo a full technical analysis "as well as operational field tests by firefighters," a requirement that would not be imposed on the Motorola models eventually selected under this contract.

As it turned out, the department wasn't ready to go to contract in February. Stogner came in for another meeting, this time with an offer to decontaminate any submerged radio at no cost to the Fire Department. Stanton wasn't interested, he said. Stogner was told that the department would not even evaluate the sample radios he'd supplied unless the hang-up button on the microphone was provided. When an April 15 date for a pre-bid conference was set, Stogner could see the writing on the wall. On March 19 he wrote a letter to city officials outside the department, including the mayor. In it, he railed about the specifications, which called for a Saber, though Motorola was the only Saber manufacturer, charging that the department "cannot tolerate" any other radio, an institutional habit for 35 years. "The exact placement of knobs, lights, displays, speaker/microphone hangers seem to be of critical importance, and not compromising," he wrote. Stogner concluded with the admonition that "it appears the state bidding laws are not applicable to radios." When he got no answer, Stogner knew his fate was sealed. He went alone to the pre-bid conference, while Motorola sent five executives and the Fire Department four.

Stogner laid out the catch-22 in the specs: Ericsson had to modify the microphone button on its radio, for example, but it also had to produce three institutional users of the "exact radio" it was offering. As he'd said in the March letter, "If you comply one way you are disqualified the other way." Stanton agreed to waive this transparent device designed to make Motorola the only bidder, since the spec replicated the features already on a Saber. But he explained that this clause had been inserted to make sure the Fire Department wouldn't become a "testing ground for untested/modified radios." Robert Scott, the department's contracting officer, who also attended the conference, later wrote that the three-user requirement ensured that the department wouldn't be "a testing ground for an unproven radio."

In the end, it was Scott who pressed Stogner about a light mounted on the top of the radio that would warn firefighters if the batteries were low. Scott described it as "a safety issue that should not be compromised." Stogner said he had no radio with such a light, but his models did have a beeping audio warning. "This is better," he declared. "They can't see the light under their coats." But the spec called for a light, so an official from the Department of Citywide Administrative Services, which supervises all city contracting, ended the review and found Ericsson noncompliant.

Two days later, the Fire Department posted a notice that it had begun sole-source negotiations with Motorola. It could do so only because Ericsson had been eliminated, and no other potential bidders had attended the pre-bid conference. A week after the conference, the department posted a list of the five radios it had selected for the Motorola contract and gave other vendors a scant five days to offer an equivalent before the window closed. No one did. The department sent the Motorola procurement to the Mayor's Office of Contracts for approval in November 1997, shortly after Giuliani was reelected. Years later, Von Essen would contend in his book that he was "advised by the mayor's office" not to put the contract out to a full bid. "We believed we were covered under our existing contract" with Motorola, Von Essen wrote, "and didn't need it" to bid it.

In addition to the contracting process, the department also had to make submissions to the budget office to get approval to fund the acquisition. Oblivious to appearances and unbeknownst to competitors like Ericsson, the Fire Department was actually relying on Motorola to structure its submission to the budget office. An October 3, 1995, internal department memo stated that the company was advising fire officials on what they should spend on Motorola's own products. "Motorola is currently finalizing a detailed technical/administrative proposal with cost estimates and procurement plan," the officials wrote.

Under the heading "Acceptance of Motorola Proposal," the memo said the company estimated the total cost at $67.8 million over five years, including 3,300 firefighter portables, EMS radios, and an infrastructure of base stations and receivers. The department sent the budget office a $68 million budget request on the same day, most of it tied to the infrastructure. By the time Steven Gregory wrote his 1997 memo, inflation had brought the estimate up to $70 million. In January 1998, the budget office approved $20.3 million for the first phase of the project—"the procurement of UHF portable radios."

Despite that pricey allocation, the Fire Department contract authorized the purchase of only 750 radios, with a paltry $963,000 in capital funds budgeted each year until 2000. A circular put out by Von Essen to the entire department said simply that "the mayor" had personally "authorized funding for the replacement of all handie-talkies and apparatus radio equipment for Fire and EMS." But all the contract did was permit the purchase of radios to upgrade those that were aging out.

The contract listed three radio options for the fire use and two for EMS, though the department was clearly free to select a radio for both of its branch operations. Contrary to the budget office and FDNY memos about a new system, the contract listed only one UHF radio among the three Fire Department options, and it was a cheaper and older version of the Astro Saber model than the one EMS was already using. Incredibly, the Saber 1 appeared on the Fire Department list—the same VHF radio it had been dumping on in repeated memos for years.

The absence of a battery light had disqualified Ericsson, but a Saber that was branded obsolete by the department eight years earlier made it into the contract. It was clear that from the moment the contract was issued, the Fire Department had no intention of buying any of the radios for fire use. It had been crafted in such small dollar amounts, featuring the same models that EMS and the Fire Department were already using, that those reviewing it in the comptroller's office or elsewhere thought it was a routine replacement deal.

Two months after it was signed, Motorola discontinued the Saber 1, removing it from the contract list and offering the department the option of substituting another model, as the agreement provided. The department sat on its hands. It had already taken nearly five months to sign the agreement after Motorola signed it on October 1, 1997; now it was taking no action when the contractor eliminated one of three firefighter options. This wasn't the usual Fire Department doldrums; this was a strategy.

After eight years of memos and maneuvers to get the contract and budget approval it now had, the department bought no new radios in fiscal year 1998 and none in fiscal year 1999, which ended June 30, 1999. With the clock ticking on the five-year waiver and only a year to go on its three-year contract, the same officials who'd been denouncing their "compromised" system did nothing. With a formal budget memo in 1999 declaring that internal communications between fire and EMS units were "virtually nonexistent," the department didn't buy a single radio that would have put the two wings on the same frequency. Officials had presumably put the models they favored into the five-option agreement they had negotiated and signed. Why were they stalling?

They were waiting for the next big thing.

In January 1999, Motorola announced that it was developing the XTS 3500, prompted in part by a memo of understanding it had signed with the National Security Agency, which was intrigued by the notion of a powerful new digital radio. Motorola was footing the research and development costs, and the NSA made no commitment to buy. The company saw the radio as a prototype it could market—especially if NSA became a lead buyer—to law enforcement agencies across the country. The radio also met the Fire Department's need for a digital UHF radio in the 400-megahertz range. Having begun development of its new blockbuster product in the fall of 1998, Motorola wanted America's premier fire department as a market-setting early customer, and the Fire Department wanted a radio that put it in the technological vanguard, outdistancing its rival, the police. When NSA dropped out of the picture, it underlined the importance to Motorola of completing the Fire Department sale.

Six weeks after Motorola announced the 3500 as a product line in development, it wrote the department again, revealing that it was discontinuing another radio on the contract list, a UHF model designated for EMS. Motorola was recommending that the department substitute the 3500 for the discontinued radio. Though it had never responded to Motorola's suggested substitutions when the Saber 1 was discontinued in 1998, the Fire Department replied to the second discontinuation in six days, asking for samples of the suggested substitute. Motorola finally sent the samples in May, not revealing that they were preproduction models. The company had offered a substitute model it had yet to develop or manufacture, a possible violation of the agreement. Instead of complaining, the FDNY salivated. It began putting the 3500s through durability and other internal tests. All the rhetoric about Ericsson trying to market an "unproven" radio, as department officials put it, was quickly forgotten, as was its supposed three-user standard.

The 3500s did not appear on Motorola's sales list until September 16. That didn't leave the department much time to test. It was still concerned about the expiration of the five-year FCC waiver in March of 2000. The question posed in marginal notes on a Fire Department budget document spelling out the timetable of the radio rollout was: "Not operational until year 2000, waiver is 5 years from what start date?" A department that had dallied for almost the entire waiver period suddenly found itself in a rush. Two months after the radio was listed for sale, the FDNY agreed to substitute the 3500 on its contract, and, for the first time since this protracted process began, it actually bought 150 radios for nearly a half million dollars. In December, it ordered 2,700 XTS 3500s, though neither the department nor Motorola had field-tested the digital UHF version of the radio in any simulated or real firefighting conditions. They were delivered in January 2000, shortly before the expiration of the FCC waiver.

The department never ordered a single radio from the list of five options that appeared in its Motorola contract. It ignored a 1998 contract that authorized the purchase of only 750 radios per year until 2000 and passed purchase orders through the system for dramatically more radios—at dramatically higher costs—without anyone raising an objection. And despite department statements in a 1997 formal review report on the contract that it would "always evaluate new technologies and products as they become available," it accepted this totally new model without looking at any other vendor's newest products. That same review report had acknowledged that "several manufacturers" were developing new digitals that might meet its needs, but the department made its overnight choice without considering any of them.

There was one more wrinkle in the works. In February 2000, as department technicians tested the 3500s in their shop, they suddenly discovered that they could not pass the submersibility test. The department had ordered and received 2,700 never-before-used, Motorola models before bothering to subject it to sufficient wet tests to determine whether it met the department's standards. The first hurdle thrown in the way of Ericsson back in 1996 was apparently not as intensely applied to Motorola. In March, the Fire Department shipped all the radios back to Motorola, which conceded that there was "a slow leak at the bottom of the universal connector." By October, the 2,700 were back, modified and dry enough to meet the standard. Even while the wet testing and modifications dragged on, the department kept ordering more 3500s, pushing the total up to 3,818 and the price to $14 million. The contract had said radio purchases were "not to exceed $2,888,910." The director of Giuliani's contracting office later attributed that language to a "clerical error."

LUKE HEALY HATES basement fires. A firefighter for 10 years, with a brother working out of a Harlem firehouse and three small kids at home, Healy was assigned to Engine Company 305 in the Richmond Hills section of Queens on March 19, 2001, responding to a 10-75 alarm. A 10-75 summons two ladder and three engine companies, and it usually means there's a fire to fight. His company was the second one due at the fire, and when they got there, they opened the back door to the basement and felt the heat warm the late winter air.

Healy was carrying a new piece of equipment, the XTS 3500R, built in Plantation, Florida, by Motorola and on the job at every firehouse in New York for its fifth day of work. At $3,154, it was by far the most expensive radio ever bought by the Fire Department, and it looked it, being both fancy and high-tech. All Healy had done with it was to "chat with other guys from the company at the firehouse just for basic communication," trying to get comfortable with a radio as different from the old Saber as any on the market. "A weird thing is we were out on the rig and I remember calling to the chauffeur on the radio and he never answered me. I thought 'no big deal.' "

In a residential neighborhood where a three-story building is a tower, the greatest danger for a firefighter is going under a house. "Basement fires are tough," Healy says. "The problem is getting down there because all the heat is coming right up the stairs. You've got to basically jump down those stairs with the hose line, and you've got to stay low at the same time. There have been times when I was at basement jobs and we got pulled out because it was too hot.

"The first thing I saw when I got down into the basement was this big red glow. So I hit it right away with the hose line and the next thing you know it went out. We're moving in more and more, squirting the hose around and, if there's any fire left, hopefully we're hitting it. It was just getting smokier and smokier and hotter and hotter and we didn't know what was going on. Then my vibe starts going off. That's something in your mask that tells you you're running out of air. I ignored it for a couple of minutes, because I figured this is a basement job, we'll be out of here in no time. Well, like I said, the smoke wasn't going away and it felt like it was getting hotter, so I told my captain 'Hey, I'm running out of air.' He said 'Okay, leave.'

"So I left the line, following it out, and all of a sudden it felt like the line was underneath something. Whatever it was, it wasn't moving, and I couldn't feel around this thing. I couldn't see anything. So I was saying to myself, 'Okay, if I'm where I think I am, there shouldn't be a wall behind me.' I was very confused at the time and my vibe alert was still going off, and I'm realizing that 'You're running out of air real quick, so basically your only option is to get on the radio.' My first thought was, should I call an urgent or a mayday? I decided, 'You better call a mayday because you don't know how much air you've got left.' So I called the mayday and I heard nothing. I thought maybe I rushed it and it came out garbled, so I gave it a second time and all I heard was static. I gave it several times and I'm waiting for a response and every couple of seconds I was trying again and then I realized my vibe alert had stopped going off.

"I was sucking the face piece because I was out of air. Basically it feels like you're suffocating when that happens. I ripped off my mask and I figured let me run for where I think the door is and I started running and I screamed for help. It turned out my captain was standing right next to me. He'd come back into the basement to get me. He could see me, but I couldn't see him, so as I screamed, he grabbed me and pulled me right to the door. I was coughing up a lung; that's what it felt like. Six or seven guys were helping me get out of the gear because they could see how exhausted I was. They put me in a basket, put me in an ambulance, and the ambulance took me to the hospital."

Healy's maydays made it into a report that Deputy Chief Michael Weinlein faxed into a citywide tour commander at 9:30 that night. The incident commander at the fire, Weinlein said he had "never had a fire where this many transmissions weren't received," including some that "stopped in mid sentence." He reported that the captain who rescued Healy hadn't heard the mayday on his radio, "but heard the member giving it in the basement." Weinlein said he did hear one of the maydays and radioed to see who transmitted it, but got no answer. Weinlein concluded that it was "very disturbing."

The next morning, four top chiefs went to the Fire Academy to test the radios. With thousands of these radios out all over the city, the only field-testing done on them was going on after they were put in service at real fires—by firefighters like Luke Healy. The department had spent a year making sure that the radios could swim. It had shown a competitor the door because he didn't have a light on top that dimmed in tune with the battery. But no one had bothered to determine whether the radios could actually convey life-threatening fear.

The four chiefs were so alarmed at the performance of the 3500s during a preliminary simulation that they rushed back to the Metrotech headquarters and reported their findings. Within 24 hours of Healy's scare, Von Essen decided that the grand experiment was over.

The digitals were on their way back to the department, and the tired old Sabers were back on the job. "That was the first time I realized that it wasn't something I did, but the radios that were wrong," Healy said later, vindicated by the recall.

A bitter Von Essen held a press conference and spoke more about Healy than he did about the radio. "In this situation, a guy could just follow the hose line out," the former leader of Healy's union and current commissioner said. "That person was probably never in any real danger. He probably just panicked." Healy read Von Essen's quotes in the newspapers and said to himself, "I wish he was in the basement with me. I wonder if he would have said the same thing if he was in the same situation." But no one asked Healy what happened, including Von Essen. The fire commissioner launched what he said was an internal investigation of the radio, and his investigators never talked to the firefighter at the center of the incident. Instead, they summoned Healy's captain downtown and two of Healy's former commanding officers. One captain got a knock on his door at 7 A.M. to notify him to appear for questioning.

"The questions revolved around the general deportment of the firefighter," Pete Gorman, the president of the fire officers union said. "What was Healy like? How did he get along with his peers? They had nothing to do with the investigation of digital radios." Gorman said the investigators read the captain who rescued Healy his rights. The only new information that Von Essen announced he'd learned from the probe was that were only five maydays, not the seven reported in the newspapers.

Three weeks after the Healy incident, Von Essen testified at a City Council hearing about the radios. Ordinarily as likable a man as anyone in the Giuliani inner circle, he came to the hearing with a Rudy-size chip on his shoulder. Since four chiefs had validated Healy's experience and unanimously recommended the withdrawal of the radios, and since they'd been withdrawn overnight, what was the point of continuing to hammer away at this firefighter? Yet Von Essen suggested that he wasn't "so sure" a nozzle man like Healy, handling the hose, should have a radio, and that he was too "quick to transmit."

"I am told he was only 10 feet in," the commissioner contended. "I am told there was not an unusual amount of obstructions. It was a wide-open cellar door. The captain said there was not a tremendous amount of smoke or heat, but this firefighter became disoriented, for whatever reason, gave a mayday. We have determined that not everybody heard a mayday. We have determined that a unit several miles away heard it. That is because of this new thing we got that had double the power that any of our radios have had." Pressed by the Republican leader of the council, Jim Oddo, about why this "particular problem didn't surface" in testing before the Healy incident, Von Essen replied, "Well, doesn't that make you wonder about how serious the problem was when it did first surface?"

No, said Oddo, it made him wonder why there was no pilot program, no testing in the academy, or no field-testing before the radio wound up in Healy's hands. "I think we have admitted that we did not do field-testing," a more subdued Von Essen replied, promising to do it now. "We will bring in firefighter Healy," he also said. It was one of many promises about the radios that Von Essen never kept. "No one from the city ever called me," Healy said in his only interview years after the incident. "I never got a call from Von Essen. I never got a call from the mayor, nothing. No one asked me if I was all right. It's also curious I wasn't called down there by the department to testify." The investigation of the Healy incident ended without anyone questioning Healy.

As tough as Von Essen was on Healy at the hearings, he rigidly defended the radio, with Tom Fitzpatrick, the deputy who would go with him to Giuliani Partners, chiming in as well. "We are not sure yet that the radio didn't operate properly," said Fitzpatrick about the unheard maydays, leaving it unclear who or what failed. "We have not said that this radio does not do what we expected it to do," added Von Essen. "We don't have any problem with Motorola's product." His explanation for why the department had already withdrawn a radio from the field that it believed worked fine was that the problem was "the way we presented it to the field."

Von Essen admitted "real-life testing should have been done up front." He said the department's protocol, in place since the early '80s, requiring that the firefighter and fire officer unions be consulted before new safety equipment is put out, wasn't adhered to—the very protocol that Von Essen as union president had frequently invoked. But his defense of the radio itself was indistinguishable from that of Motorola vice president Ken Denslow, who also testified.

Denslow returned the favor—not disclosing that the 3500 had an orange mayday button on it that would have told Healy's captain which firefighter was in trouble and what his position was. Motorola never revealed that the FDNY had asked the company not to turn this button on, and that, as a result, the thousands of radios sent to firehouses had dead mayday buttons on top. The department regarded it as a new feature it hadn't asked for, and planned to activate and introduce it down the road. Since Von Essen was accepting responsibility for a training screwup and defending the radio, Denslow never invoked the button defense, which would have helped the company but embarrassed the department.

Similarly, Von Essen didn't say anything about Motorola's obligation to field-test a totally new product line. He would later write in his book that when he erroneously told reporters that 3500s were already in place in Chicago and Boston, he was merely "passing on incorrect information from the manufacturer." But when asked about the false information at the hearing, he blamed it on an unidentified staff person, saying he refused to "single out anybody" on his "team."

Von Essen later boasted in the book that "nothing came of the investigations" because "nothing could have," since "we were guilty of nothing more than a few mistakes." In his list of "nothings," he left one out. The Department of Investigations, which was legally charged with probing the allegations, investigated nothing.

WHEN THE DEPARTMENT went back to the antiquated VHF radios again, it was, of course, turning its back on the now six-year-old promise to the FCC to shift to the interoperable UHF frequencies granted under the spectrum waiver. Fortunately, the Police Department had so aggressively utilized the spectrum that it appeared unlikely that the

FCC would move to take it away. In addition, no broadcaster had applied for control of the frequencies. Until one did, the FCC would allow the 12 public agencies to remain as "holdover tenants," occupying the band beyond the waiver. Still, the delays did represent a violation of the agreement. The Fire Department had been granted 48 UHF channels as part of the Channel 16 expansion, and it was still on VHF. Six years after Giuliani announced his grand covenant with the FCC, firefighters still couldn't talk to cops or paramedics. There was no press conference, or even press notice, about that development.

Giuliani, in fact, left office without taking any formal action to safeguard the Channel 16 frequencies in his final two years. Three months after he was gone, Bloomberg officials asked the FCC to make the spectrum grant permanent. At that hearing, Agostino Cangemi, the deputy commissioner of the Department of Information Technology and Telecommunications, called Channel 16 "our most critical post–September 11 public safety spectrum need." He cited the police and EMS use of the UHF frequencies. But seven years after the waiver was granted, he still had to talk prospectively about the Fire Department, saying it was "in the process of deploying thousands of fire-ground radios" on the band.

Nonetheless, when Giuliani made his May 2004 television appearances after the radio chants dogged him at the 9/11 Commission hearings, he repeatedly raised spectrum as both the problem and the solution in emergency communication. "What about the radios?" he asked on CNN, echoing the audience chant. "Some worked and some didn't work. The solution has to be creating a bandwidth that is dedicated solely to emergency services so that if I arrive there and the fire commissioner arrives there and the police commissioner and everyone else, we can communicate on a dedicated bandwidth." Giuliani did not mention that his fire and police departments—the only two agencies whose commissioners reported directly to him rather than through a deputy mayor—were on different frequencies on 9/11 because the FDNY couldn't figure out how to get on the dedicated bandwidth given them by the FCC (and because OEM couldn't get them to use the 800-megahertz channel).

Giuliani's other post-9/11 radio position was that the communications failures were technological, not a shortcoming of his own administration. Richard Ben-Veniste asked Giuliani the only question at the May hearing involving radios and focused on the importance of police/fire communications. Having worked with Giuliani as a prosecutor in the '70s, Ben-Veniste got personal: "Given the fact that you were no shrinking violet, and given the fact that the differences in the radios made it obvious that there could not be easy inter-agency communication, what barrier was there that prevented you from ordering standardization?"

"No barrier," the mayor replied. "The technology, and that's why there isn't standardization today. They should have radios that are interoperable, so that in an emergency, both of them could be switched onto the same channel. Those radios do not exist today." Giuliani's answer defied fact. The U.S. Conference of Mayors had just completed a study of 192 cities that found that 77 percent had radios that were interoperable across police and fire departments.

Countless documents from Giuliani's budget, contracting, fire, purchasing, information, and emergency management offices justified tens of millions in expenditures over the years by citing interoperability as a central purpose. Way back in 1995, the city's information agency reported to Giuliani that the FCC's waiver "assures interoperability among the police department, fire department, Transit and Emergency Medical Service." But instead of getting all the emergency services on the same channel, the administration squandered opportunity after opportunity to achieve that goal. It did it on behalf of a Fire Department radio deal—with the mayor's office authorizing the expenditures and advising on the no-bid process—that was as hapless as it was contrived.

ON TOP OF all the heat Tom Von Essen was getting from the City Council, Comptroller Alan Hevesi was also on his case. He'd already stopped payment to Motorola, stating that "the FDNY broke the contracting rules." He cited "misrepresentations by FDNY staff" who purchased the radios, a survey of nine cities finding that none used the 3500s, and the "inappropriate use of purchase orders" to move from 750 to 3,818 radios. He also highlighted 12 field reports written by fire officers faulting the radios during their only week in the field, some of them predating the Healy incident. Those reports directly contradicted Von Essen, who claimed that Healy's missed mayday was the first warning of any problem.

One report—written by Chief William Blaich, the same officer who fearlessly took on the fuel distribution system at 7 WTC—revealed how badly the 3500s had performed at a trade center fire the day they were first put in service. There was, he noted, "no communication from the concourse to the street." The next day the "battery went dead at an operation"—so much for the missing battery light that killed GE/Ericsson. Another battalion chief reported, "I feel like I'm in a Godzilla movie. The lips are moving and the voices are coming afterwards."

The comptroller also subpoenaed Captain John Joyce, a supervisor at the Fire Academy, who persuaded the communications director, Steven Gregory, to give him 20 radios two months before they were put in service. Joyce used the radios with 482 trainees over a six-week period on his own—not as a field-test but in an effort to familiarize firefighters with equipment that would soon arrive in their firehouses. Asked under oath if he had an opinion as to the effectiveness of the radio, he testified, "They suck." Asked if he would feel safe if firefighters were using them in the field, Joyce said, "No." Asked if he'd personally observed situations where firefighters were standing next to each other and unable to communicate on the radio, Joyce replied, "Every time we used the radio."

Gregory instructed him not to put any assessment in writing and, when Joyce called Gregory and told him what was happening, the assistant commissioner said, "We know about the problems." According to Joyce, Gregory refused to send up someone from the tactical unit at Joyce's invitation to see what "we were encountering." Then on February 9, he got a phone call at home telling him that the 20 radios he had "were needed at the commissioner's office ASAP." They wanted the combination to get into the locker where the radios were. When they were delivered downtown, Von Essen's executive assistant signed the receipt for them. With that testimony, any question about whether the highest levels of the department hadn't knowingly put a troubled, untested, new radio out for service went by the boards. Instead of responding to the problems that Joyce found, the department reclaimed his radios a month before it put them in every firehouse.

In May, Hevesi cited the testimony briefly in a letter to Motorola CEO Tom Galvin without naming Joyce or laying out any of the captain's evidence of Gregory's and Von Essen's resistant roles. He also mentioned the "near tragic" Healy case. He assailed Motorola's "rush to market without any significant field testing" and said it "raised serious questions about the company's motivation in so aggressively pushing this particular product on the FDNY." He accused the company of disingenuously refusing to disclose which other governmental agencies had purchased 3500s because of supposed "confidentiality restrictions," though it had issued 40 press releases in 1999 announcing government sales of other equipment, even when the customer was the FBI.

Unaware that he was echoing the testing-ground language the Fire Department had used to disparage GE/Ericsson all those years earlier, Hevesi charged that Motorola had "allowed the FDNY to unknowingly become a large scale testing ground for the XTS 3500 radio to the possible detriment of NYC firefighters." Hevesi asked Motorola to recall the radios. Opposed by the department that had ordered and withdrawn them, it was a doomed request. Motorola took nearly two months to even reply. In a June 27, 2001, letter, Denslow denied Hevesi's charges but provided few specifics. He suggested that it was the Fire Department that was aggressively pursuing the radio, not Motorola pursuing the department. He said that fire officials had "on more than one occasion asked if Motorola could accelerate its schedule," implicitly confirming the earlier indications of the department's conscious avoidance of the radios actually listed in the contract.

THE FIREFIGHTERS HAD been asked to leap a huge technological hurdle when they were abruptly handed the new radios that operated so differently than the ones they had grown accustomed to over many years. At the City Council hearing, Von Essen raised the possibility that the adjustment could be eased if the department reprogrammed the computers that controlled the 3500s so that the transmissions occurred in the same analog voice as in the old Sabers, rather than the distended and millisecond-delayed digitized voice that they had in the initial rollout. "Motorola is working with us," he said, "trying to help us make any program changes we want to make." The 3500s were perfectly capable of speaking either way, but the department had opted to switch to the digital voice, even though it was already asking firefighters to adjust to a wholly new UHF technology.

Von Essen announced at the hearing that he'd appointed a committee of outside experts to advise him. One of them, Jerry Hauer, also testified. Hauer, who'd been out of the administration for more than a year, said the department had tried to do too much at once. "You were changing radios, you were changing bands and you were also changing technology," he said. "Putting it out there as an analog until people have gotten used to the radio and the new technology and evolving to digital would have been a stepped process."

If the new radios weren't reconfigured and returned to the field, the old ones, discontinued by Motorola in 1998, would remain. Von Essen made it clear what he thought about that possibility. "We are going to test and document the old radio before we do anything about the new radio. Because nobody has mentioned these past two weeks all the problems we have with the old radio." He reiterated all the familiar complaints, deriding a system "we have been using for 35 years" and observing prophetically that the department would be "able to communicate with the NYPD with the new radios when we get them back into the field," in contrast with VHF Sabers. A letter from a Motorola lawyer to city officials contrasted the "present, beaten up, 18-year-old analog radios" with the new ones. "No one who truly cares about the safety of FDNY personnel should want to delay the deployment of the new radios," contended Scott Mollen, the company's New York counsel.

Yet Von Essen refused to answer questions from council members about when he expected to finally replace the old radios, presumably with reconfigured 3500s. "I don't even want to give you a time line. I will not rush this process," he said. He mentioned his panel of experts. He mentioned training. He seemed more interested in using the dysfunction of the old radios as a rhetorical defense than an imperative for action.

One of the reasons surfaced briefly when Peter Vallone, the council speaker, pressed Von Essen about including the unions on his review committee. "I don't have any problem having the unions involved," said the commissioner. But, in fact, he did have a big problem with that. Von Essen and Giuliani himself were publicly attributing the whole brouhaha over the radios to the unions, not the Fire Department's own failings. It was, said Von Essen, "brought out by the angst we're going through with the unions." Giuliani said the union was "exploiting the controversy because of past disciplinary decisions" by Von Essen. If the commissioner had to include the unions in his decision-making process, he was unlikely to really try to make a decision.

The one time the task force met in Von Essen's conference room in June, he invited representatives from the two firefighter unions to join the group, as Council Speaker Vallone had urged. But the union for fire officers wanted a board member who sat on their safety committee to attend and, though a respected chief in the department, he was barred at the Metrotech entrance. "The Fire Department will never dictate to the union who will attend meetings," charged union leader Pete Gorman. The other union's representative walked out in protest.[5] Any Von Essen effort to put the radios back without the unions was a breach of protocol and his public promise. But including them—at least on their terms—was too much for Von Essen to swallow. He couldn't live with them and he couldn't live without them.

Von Essen stood dangerously still, apparently unmoved by his own and everyone else's description of just how bad the old radios really were. Forgotten was the pledge by Chief of Operations Daniel Nigro on the day the 3500s were withdrawn that "the operational issues with the XTS 3500 will be addressed in the very near future and they will be returned to service shortly." Hauer said Von Essen's task force met only once more after the council hearing and did no report. "We clearly identified the problem and agreed to convert the digitals to analogs," he recalled. "I don't know why they weren't sent back. It went slowly. I never heard anything else."

It took until August before the department began limited tests of the new and old radios at the fire academy. Five precious months had passed since the old radios Von Essen had condemned were returned to firehouses. He had only five more months to go on the job. He assigned a chief to begin tactical training with the radio in analog and digital mode and, according to the department's spokesman, the training was scheduled to start in September.

By then, everybody in the administration, including the mayor, was busy planning their new lives. Vallone and Hevesi were pitted against each other and two other Democrats in a contentious primary. Giuliani had dismissed Hevesi's critique of the radios, attributing it to the comptroller's desire to "sway the FDNY unions into backing him," a convenient alternative to considering the accuracy of any of the charges. "The fire department needed new radios," the mayor also said back then, defending the purchase. But months later, he appeared unconcerned when the department still didn't have them.

In Von Essen's private 2004 testimony before the 9/11 Commission, that gnawing conscience of his spoke again, when he raised the question of whether the 3500s, if not withdrawn, would have operated better on 9/11 than the Sabers. "Digital communications were stronger, could go further. This ended up being a problem in the field. I believe with proper training it could have been overcome. I have asked people since then if digital radios would have made a difference on September 11, and they are not sure if it [sic] would have. I have talked to people at Motorola, and they say they are not sure that it would have been better."

As genuine as Von Essen's belated search for truth might be, he posed the wrong question. If the digital radios had been adjusted to speak in the familiar analog mode when they were first issued or reconfigured quickly after the recall, would they have been better than radios his own department branded as "the oldest, obsolete and most unreliable portables?" Or if the Fire Department had ordered the analog Astro Sabers that were on its contract and were used by the police on 9/11, instead of the 3500s, would they have been better? Or if the 3500s had been urgently returned to the field with training that clearly explained the differences of a digital voice, would they have been better?

The answers are axiomatic, especially when the repeater issue is added to the mix. The Port Authority installed repeaters, or amplifiers, after 1993 to improve the range of the old analog radios in the towers. Von Essen and other Fire Department officials have tried to blame the 9/11 communications breakdown on the fact that the North Tower repeater did not work that morning. There is no doubt that the chiefs tried to start it and couldn't—the only question is whether it failed or they did. But the authority had also installed another repeater in 2001, at the request of the Fire Department, to heighten the reach and power of the new 3500s. And unlike the old repeater, the new one did not have to be turned on. It was on all the time.

Had the 3500s been in service with adequate training and an automatic repeater, would they have been more useful than the ancient Sabers minus a repeater? Even Von Essen said the digitals offered better penetration. They were so powerful that Healy's mayday was heard miles away without amplification. They had a channel that would have directly connected the Fire Department and the police. While there are legitimate questions about how well the flawed 3500s might have worked, or how GE/Ericsson's model would have performed, can Von Essen possibly believe that the Sabers derided by the department's top technical people in a 1990 memo were the preferable equipment?

Von Essen can find his answer in every firehouse now. The Bloomberg administration returned the 3500s to the field, starting in Staten Island in February 2003. They were reconfigured to communicate in an analog voice. The mayday button was activated. They were introduced slowly and with training. "It's much better now, that's for sure," said Luke Healy, praising the mayday button in particular. "If the radios were like they are now, in the analog mode, I don't think there would have been a problem" at that Queens fire. Healy's company was kept in a staging area for the 102 minutes of 9/11, though he made it to the WTC later that afternoon. Had the analog 3500s been re-issued before then, says the firefighter Von Essen never spoke to, "I would say there definitely would have been more lives saved."

If Von Essen has, since 9/11, proven to be willing to consider what his successor, Nick Scoppetta, calls "the sins of the past," he has also been willing to play retroactive radio hardball. In his 2002 book, *Strong of Heart*, Von Essen blamed the unions and Hevesi for the fact that firefighters still had the same old radios when disaster struck, saying their actions were "very hurtful to firefighters and officers." He wrote that "the radios we pulled were a better product than what we had," but "the politics of the situation required us to hold back their implementation," suggesting that the failure to reissue the 3500s was because of some unspecified political obstacle. He told a reporter much the same thing, conceding that the digitals were "better radios in the analog mode," but that "because of politics, we never got them back out fast enough."

What kind of "politics" stalemates a lame-duck and term-limited administration on a matter of a life-and-death technological upgrade? Was it just that neither he nor the mayor wanted to appear to be giving in to Hevesi, Gorman, and any other "enemy" that had pushed the radio agenda? Was Giuliani's attitude on radios in 2001 simply "Fuck Hevesi," as one aide depicted it? What other kind of "politics" could there be in a City Hall whose master was already designing a consulting portfolio?

Von Essen has never explained. But having selected his scapegoats, and oblivious to jarring warnings about the obsolescence of the radios returned to the field, he was content in the months after the Healy uproar to slump serenely, together with his mayor, toward the cataclysm of incommunication that killed firefighters on September 11.

THE TECHNOLOGY AVERSION at the highest levels of the Giuliani administration was legendary. The mayor never carried a radio or a cell phone. A top deputy with him on September 11 says, "Someone gave him a cell phone that morning." The phone, apparently a Nextel from

OEM, wound up in the Smithsonian, and a museum spokesman said it was handed to Giuliani as he walked north up Church or Broadway. A television reporter walking with him, Andrew Kirtzman, dialed his station on his own cell phone and gave it to the mayor as they walked uptown. Giuliani wanted to talk to Kirtzman's audience and instruct everyone to walk north, but the cell failed.

"The bodyguard usually had the cell," Giuliani's deputy explained. "Denny Young, who was so often with him, never had his on. He only made calls, never received them." Similarly, neither the police nor the fire commissioner usually carried radios. Indeed, when Tom Fitzpatrick, the Von Essen aide who oversaw the radio deals, was asked during his oral history interview whether he had had a Fire Department radio with him that morning, he said, "I had no radio, no. I just had my turnout coat." At one point, before the collapses, a top chief asked Fitzpatrick to reach someone. "I can't for the life of me remember who or what," Fitzpatrick recounted. He wound up jumping a wall in front of the World Financial Center, and running into a security office. "I made a phone call and came back out in the driveway," he said.

Giuliani did, however, create a new technology agency when he took office—DOITT, which stood for Department of Information Technology and Telecommunications. (It was pronounced, Nike-like, "Do it"). Ralph Balzano, the great champion of the Metrotech command center, was the first commissioner. Like Kerik, Von Essen, and Sheirer, he came into the administration straight out of the campaign. An associate of Bronx Republican leader Guy Velella, Balzano had put together a computerized voter identification system for the state GOP's "Victory '93" committee that helped Giuliani win.

Balzano was brought into City Hall and asked to help organize DOITT, which was a merger of three other units spread out over city government. The main focus of the agency from the outset was data technology, Balzano's strong suit. It also oversaw the city's cable franchises and stations. Thrown in, almost as an afterthought, was interagency radio communications, and neither Balzano nor his eventual successor, Allan Dobrin, had any background in the field. None of their top deputies did either, making radios a sideshow within a sideshow, since DOITT never got much attention at City Hall and this slice of the agency got even less. When Giuliani interviewed Balzano for the position, the radio side of the job never came up. Neither did the 1993 bombing or terrorism. As legitimate as the technology imperative was to forge the agency, Giuliani's cobbling together of its functions had nothing to do with integrating data and telecommunications as part of the city's emergency preparedness.

"Radios were not my strength," said Balzano, who ran the agency until 1998, "and I don't think it was a main concern of City Hall. The mayor never spoke to me about radios, and neither did Peter Powers or Randy Mastro, the deputy mayors I reported to. I put references to it in my regular reports to City Hall, but they never asked anything about it. No one expressed any interest." Balzano wasn't even invited to the 1995 press conference when the FCC spectrum decision on Channel 16 was announced, though it became his agency's responsibility to coordinate the multiagency compliance with the five-year waiver. "I think the consensus at City Hall was that radios were a Police Department and Fire Department issue. They were the owners of most of the radios and the issue."

To Balzano, it was all quite natural. "Rudy understood crime, so he talked about it. But he never understood telecommunications, so he didn't," said the ex-commissioner. "He didn't like to discuss items he didn't know much about."

After Dobrin replaced Balzano, it took three more years before the mayor asked him the first question about radios. Every two months, Dobrin trekked over to City Hall to cabinet meetings around a huge round table, but the question of public safety communications never came up. Dobrin got the DOITT job because he'd proven himself a jack-of-all-trades in the Giuliani years—he was deputy director of operations and of Bellevue Hospital, as well as executive director of the special education committee. By the time he succeeded Balzano, the big challenge for DOITT was getting the government more Internet-friendly.

When the Healy incident began to dominate news coverage in March or April 2001, Giuliani got Dobrin on the phone and posed his only radio question. He said he'd been talking to the Fire Department about the radios and wondered what Dobrin thought. Dobrin threw out the strengths and weaknesses of digital and analog. He said that the Fire Department had to make an operational decision about what was best for fire protection. Giuliani didn't press Dobrin to express his own opinion about the choice; he just wanted, six years after the Fire Department's decision-making process began, to understand the difference.

Giuliani called Dobrin because DOITT had played a pivotal role in advising the Fire Department about radios. In fact, the February 27, 2001, application filed with the FCC to license the Channel 16 frequencies for the 3500s was signed by the DOITT associate commissioner in charge of radios, Deborah Barell Spandorf. The daughter of the former chancellor of the State University of New York in the Cuomo administration, Spandorf had taken charge of radio policy in the Dinkins days. Her unit was considered so unimportant that when the first deputy mayor, Peter Powers, sent a trusted aide over to "case" the telecommunications operations for him, Spandorf's office went unmentioned in the three-page 1995 assessment.

Nevertheless, Spandorf's little universe within DOITT wound up handling the issues that would have the gravest long-term consequences. She was in charge of Channel 16—the best hope for finally getting police and fire on an integrated network. She was also critical in pushing the Fire Department toward a digital system that a subsequent City Council report found that her agency "had determined that the fire department should move" to the 3500s. A DOITT deputy explained why at the council hearing about the withdrawal of the radios, testifying that the agency urged the selection of the 3500s because digital radios were "consistent with the strategic direction that DOITT was going in with Channel 16"—Spandorf's project. Tom Fitzpatrick confirmed that the Fire Department's move to digital was part of "a citywide plan for all agencies to use the same radio with the same specifications so we could communicate with each other"—a reference to Channel 16.

Digitals didn't turn out to be the "requirement" that Fitzpatrick said they were, however, since the Police Department, which spearheaded Channel 16, refused to buy any. Police officials rejected high-level efforts to force them to buy the same digitals the Fire Department did. Instead, they opted in 1999 to stick with their Astro Sabers, an analog Motorola product.

In addition to the FDNY digitals, Spandorf was a key player in a multiplicity of other city deals with Motorola. With her in the lead, DOITT bought $7 million in Motorola digitals for an 800-megahertz project, the other interagency communications system it championed. Working with OEM's Jerry Hauer, DOITT's 800-megahertz system eventually involved 40 city agencies, including Fire Department chiefs. But it had little public safety utility, partly because the Fire Department rarely used it. During the Spandorf years, DOITT also put together the specifications for an expensive Channel 16 interagency system that Motorola wound up winning after 9/11—at a $75 million price tag. The full Fire Department system that DOITT pushed was projected to cost $70 million, including radios and infrastructure, and Motorola was again the designated winner.

Spandorf personally proposed a $27 million, three-year maintenance contract with Motorola in 2000. In fact, her unit flew out to Motorola's corporate headquarters in Schaumburg, Illinois, for periodic conferences. Motorola was so confident of the business relationship with her that when it had to fill out a 1996 bidder's qualification form for a DOITT contract and was required to list three of its "largest, most recent customers," Spandorf's name was first on the list, followed by the FDNY's Steven Gregory.

Warren Stogner, the GE/Ericsson salesman, said Spandorf "operated autonomously within DOITT," and that she "was always really nice to me. But she refused to let me talk to the people in her shop who were preparing the specs, which I thought Motorola did. She just never bought from me." He said she "built a fiefdom," told all the other agencies "we've decided to standardize," and then never allowed anyone but Motorola to win." In October 1997, seven months after Stogner's letter about the Fire Department contract, he wrote another letter, directly to the mayor. He complained that DOITT was "procuring a radio system without competitive bidding or allowing vendors like GE/Ericsson to offer their products."

It wasn't as if *no one* liked Ericsson's products. Stogner pointed out that it was providing the citywide system for the state police and a countywide system in neighboring Nassau County. Later, the company beat Motorola on a multiyear contract covering virtually all New York State agencies valued at nearly $2 billion, despite the fact that Motorola lobbyists included the former head of the state GOP and John O'Mara, a partner of former GOP senator Al D'Amato and a fixture in Governor Pataki's office. As successful as GE/Ericsson—which became M/A Com—was with that contract, "we finally just stopped trying to sell to the City of New York," said Stogner.

What nobody knew was that Spandorf's sister, Marian Barell, was an in-house lobbyist for Motorola working out of its 30-person, Washington-based office. Before starting at the company in 1995, Barell had her own consulting firm, and Motorola was a client. She did some telecommunications lobbying, but her title was vice president and director of global trade and her job, as she put it in an interview, was lobbying for government actions that "promoted Motorola sales overseas." The former deputy assistant to the U.S. trade representative in the Reagan and Bush I years, Spandorf's slightly younger sister retired from the company in 2003, and moved out near its Illinois headquarters.

Informed of this relationship years later, Balzano said, "She should have disclosed that and never did. There's no way to find out unless she tells you." He said she was certainly supportive of Motorola as the city's key radio vendor. "But the only company I ever heard anybody talk about in the administration was Motorola," he said. "Radios were synonymous with Motorola." Though Spandorf also never told Dobrin, he refused to answer questions about her.[6]

On October 1, 2001, the 50-year-old Spandorf was found dead in the bathtub of her East 86th Street apartment. The super in her building went into the 17th-floor apartment when the neighbor below it complained that running water had been leaking through the ceiling for a day or two. The police contacted members of the family, including her sister Marian Barell. The medical examiner did not rule the death a suicide, but Barell said it was her "understanding that the police never definitely determined the cause of death." A DOITT official close to her said he wasn't sure if it was a suicide, but said that when she was found she was "heavily, heavily medicated."

Spandorf had been depressed for years. Even in the mid-'90s, when Balzano was running the agency, Spandorf's depression kept her out of the office for long stretches of time. Dobrin had to stop staff meetings and walk her outside. DOITT officials talked to her psychiatrist and learned that she "couldn't get stabilized on the numerous medications" she was using. Described as looking like a "trim, pom pom girl" by one businessman who dealt with her, she was also known, even by those who worked with her for years, as "someone who never laughed." She'd been separated from her second husband for five years, and the divorce became final right around the time she died. A boyfriend died of cancer earlier in 2001. And 9/11 had traumatized her.

Friends at DOITT say she'd rarely been at work for weeks before 9/11, but that wasn't unusual. What was unusual was that she showed up after the attack—at the command center at Pier 92, where Giuliani established his headquarters on September 15. DOITT staff members remember her being so upset that she collapsed at the pier and "had to be hospitalized at the scene." She was put in a hospital bed on a navy ship docked at the pier, they recalled. Her personnel records indicate that on September 23, she was placed on sick leave without pay. That was the first time in her 16-year city career that she was recorded as taking a leave of absence, paid or unpaid, despite what several sources describe as protracted periods of absence.

Her 30-year-old daughter, Audrey Martin, said later, "I think the combination of the stuff that happened on 9/11, and the stuff with her boyfriend, and the stuff that was going on at the job, it led her to probably take more medication than was probably good for her. But it was all prescribed. Ultimately they ruled she had hypertension, which is something that she had for a long time. Probably the drugs she took didn't help things. But I'd spoken with her the morning that she died and she had been planning to come out to Seattle to see us." Martin had seen her mother in August, and thought she "was a mess then" but "more lucid" afterward on the phone, at least before 9/11.

Asked if her mother ever mentioned Motorola, she said, "Yeah, obviously because she worked with them and my aunt, her sister, was a big head honcho at Motorola. I know that she worked with them. I know that when I called her on 9/11, the line was jammed. Afterwards, when I talked with her about 9/11, she said that her department had been working around the clock with communication issues because it was such a mess. She had a lot of friction with her manager. That part of her job made her very stressed out."

For five years, a big part of the stress was something highly unusual in a municipal bureaucracy: a real deadline. Spandorf took the March 2000 deadline for Channel 16 so seriously that she used the FCC's short window to justify no-bid contracts for Fire Department and other agency radio infrastructure. Even though the FCC order only "encouraged" a move to digital technology, and explicitly said it was not requiring it, Spandorf interpreted the clause "reasonable efforts" as a digital mandate. So when DOITT and the Fire Department failed to meet the 2000 deadline and wound up in 2001 with digital radios that sparked the Healy firestorm, Spandorf was on the hot seat.

The 800-megahertz radio project was also her baby, and that, too, was a bust, with virtually no one using the radios that operated in that often crowded range. A City Council report would later blame the failure of fire and police officials to talk on the 800-megahertz radios on the fact that city officials—clearly including Spandorf—"had not delineated procedures for using interagency radios." The 800-megahertz and Channel 16 projects had one common objective: interoperability. Yet Hauer, who worked closely with Spandorf, had to concede in his commission testimony that, over his four years in the administration, "the interoperable radio project that DOITT had been working on continued along with slow progress." All of these issues were said to have weighed on Spandorf in the months before 9/11, and apparently after it.

The post-9/11 Spandorf question at a DOITT and city level wasn't her Motorola ties, since no one could have uncovered and regulated that. It was her years of delay and dysfunction. How could the city's central interoperability manager go on for so long with such medical and personal disabilities? Was anybody at the highest levels paying attention? Incredibly, Deborah Samuelson, the deputy counsel at DOITT who handles freedom-of-information requests, maintained that the department did not have any Spandorf files or documents covering her seven-year history with the agency. When lawyers for the families of dead firefighters sued and sought Spandorf records, they got the same mystifying answer.

Even Hauer, who unlike the technologically averse Giuliani inner circle carried a belt full of radios and cells, was so consumed by command center and incident command issues that he couldn't take on the communications mess. The agendas of his mayoral meetings referred only twice to radio issues, and once it involved his failed effort to find a way for police and fire personnel in moon suits at a hazmat incident to talk to each other. "Steve Gregory didn't want the police department talking on a fire department frequency at a hazmat incident and he blocked it," Hauer remembered. If he couldn't even get them talking to each other "mouth to mouth, chief next to chief" at these exercises, Hauer didn't believe anything would happen on the bigger issues. "I was very skeptical that interoperability was going anywhere," he said. "The police department and the fire department would never get it done." So he spent his time on the doable, or possibly doable.

Spandorf didn't have that choice. And then she woke up one morning and found out that people died because cops weren't able to talk to firefighters, chiefs didn't know how to use their only interagency radios, and firefighters still carried potluck radios as old-fashioned as typewriters. Her memorial was on a boat docked at a Long Island pier, and her sister Marian, daughter Audrey, and several DOITT friends were among the speakers. Barell was the family member who called DOITT to give them the memorial details. In an interview years later, she insisted that she "didn't even realize" that her sister "worked on Motorola business until shortly before she died." She'd come to New York as recently as the summer of 2000 and spent some time with

Spandorf, she said, but they were not in close touch. "There was nothing hostile," Barell said. "We'd just gone our separate ways."

Citing her international responsibilities at Motorola to distance herself from her sister's city business with the company, Barell didn't mention that she was now married to Motorola executive vice president Robert Barnett. A member of the Motorola board, Barnett has long run the commercial and governmental sales division, specializing in telecommunications. Barnett ran Motorola's iDEN group in the '90s, an "integrated dispatch network" that Spandorf's unit traveled to Illinois to learn about and eventually bought. Barell didn't marry Barnett until 2003, but his former wife says she began proceedings in late 1998. Divorce papers indicate that the Barnetts had been separated since at least 1999, and Barell stopped using her married name on Motorola lobbying filings in 2000. Her separation agreement indicated that she had plans to relocate long before she did in 2003—when she retired from Motorola at the age of 50 and moved out to the Chicago suburbs to marry Barnett. Neither Barnett nor Barell would answer questions about the timing of their relationship or whether Spandorf dealt with Barnett's unit, referring those issues to Motorola, which also declined to answer.[7]

After Spandorf died, DOITT hired Steven Harte to replace her. Harte was a top Motorola salesman in New York. He was the key person from Motorola who dealt with Dobrin, Spandorf, and DOITT. He described himself in a form filed with the city in early 2002 as having "conducted business" for Motorola in 2001 "with various city agencies." He'd become a permanent presence at Giuliani's command center at Pier 92 in the aftermath of 9/11, bringing in Motorola equipment and helping to set up communications there. Having made a big impression at the pier, according to one friend, he took over Spandorf's $112,800-a-year job at the start of the Bloomberg administration. Not surprisingly, over the subsequent years, Stogner found him to be a "pretty tough sale." Spandorf and Harte weren't the only decision makers with Motorola connections. Two police captains who handled radios, Ken Weinberg and Steve Ahmed, left the city and went to work for Motorola. Weinberg signed the Police Department's contract documents with the company.[8]

Von Essen, now a member of Giuliani Partners, traveled to Las Vegas in April 2002, and was featured at a Motorola conference for its dealers and franchisees. He delivered his "Leadership Lessons from 9/11" speech right after "Our Changing World" by Ken Denslow, the vice president who'd appeared at the same City Council hearings as Von Essen in the middle of the Healy incident. Von Essen speeches were listed at the time as costing $15,000 to $25,000. The theme of the conference, held at the Aladdin Resort & Spa, was "Ready 2 Respond."

Saying repeatedly what an "honor" it was for him to be addressing what he called a "Big M" audience, Von Essen heaped praise on the company for "the support that came when Motorola came to the command center" after 9/11, supplying vital equipment. "Knowing how the City of New York is," he added, to laughter, "I'm sure that Ken"—an apparent reference to Denslow—"is still waiting for some checks to come." Thanks, he ended, "from the firefighters especially for all that you did in making it possible to get through the worst thing I hope that the firefighters and all the people of New York will ever go through."

He and Giuliani had one other post-9/11 connection to Motorola. One of Giuliani Partners' earliest and biggest clients was Nextel, a company with longstanding connections to Motorola. The largest single shareholder in Nextel for most of its history, Motorola licenses and equipment jump-started and sustained the company. It manufactured practically everything Nextel sold. Giuliani, Von Essen, Kerik, and Sheirer were retained in May 2002 to drum up public safety community support for FCC approval of what was derided by critics as a huge Nextel spectrum grab. Motorola was slated to be paid hundreds of millions of dollars by Nextel to complete the rebanding necessitated by the spectrum changes. Obviously, Nextel's retention of Giuliani Partners was unconnected to the city's Motorola contracts, but the irony of the Four Horsemen, within months of 9/11, charging around the country gathering public safety support for a deal that benefited Motorola is breathtaking. It's also one more indicator of the common wavelength they were all on.

As compromised as the city's relationship with Motorola was in the Giuliani era, it was not just a Giuliani phenomenon. The company had sunk its roots so deeply inside city government that while mayors came and went, the relationship remained. Jack Mohan, who was brought into the Fire Department in the 1970s "to resurrect the dispatching system," as he puts it, remembers a confrontation he had with the department's communications director. "As I was writing the spec," Mohan recalls, "I discovered that there already was a spec that communications had written, and buried way deep in this specification, at the bottom of a page at most, was this spec for a dispatching system for Motorola. I knew Motorola wrote that spec for them. They were using Motorola to write all their specs. They were very much in bed with them, dating back to those days. They were shoveling this whole thing to Motorola." Mohan, who now owns his own software company, says that Steve Gregory was then "the right hand man" of the communications director he confronted.

The problem was that no one at the highest levels of the Giuliani administration, including the mayor, noticed. It wasn't that this intertwine was below the radar; it's that there was no radar. Stogner's 1997 letters to the mayor sounded no alarm. Even the firestorm of March 2001 was dismissed as if a perceived political motive was all the "apolitical" Giuliani needed to justify ignoring credible charges. John Lehman, the 9/11 commission member whose Boy Scout parallel provoked such ire, used equally tough language in an interview about the radios, calling it a "contracting mess in City Hall with scandals over procurement" and saying that the highest levels of the Giuliani administration should have "overruled the sweetheart deals." He said that "the approach was if we buy the right box" at each agency, everything will be fine, instead of organizing "a small signal corps," like the military, that would be in charge of communications in a major emergency.

"If you as a city don't process your requirements correctly," Stogner added, referring to the closed bid process, "there's no reason for your vendor to do the right thing for you. The problem with having only one vendor is you have to do what he tells you. You're going to buy mine no matter what I do, so there's no reason for them to treat you right."

THE DAY OF reckoning for firefighter families like Al and Sally Regenhard was March 4, 2004. That's when 15 to 20 families, 12 of whom were still named plaintiffs in the case of Lucy Virgilio versus the City of New York and Motorola, appeared in the small but always busy courtroom of U.S. District Court Judge Alvin Hellerstein, who was presiding over virtually all 9/11 litigation. Hellerstein was to decide the fate that day of the Virgilio case, which charged that firefighters had died "unnecessarily" because "no substantial modification was made to FDNY radios between February 26, 1993 and September 11, 2001," the dates of the two terrorist attacks.

All of the families had already faced a deadline they dreaded. By January 22, 2004, they had to sign up to participate in the Victim Compensation Fund or forfeit whatever sum Kenneth Feinberg, the fund's special master, might assign to their loss. Feinberg paid an average of $2 million per family to survivors of the 2,880 people killed in the attack, though it varied widely based on the earnings of the deceased. Ninety-eight percent of the survivors took Feinberg's money. Those who did sacrificed their right, by law, to sue everyone but bin Laden. At least, that's the way the courts ultimately interpreted the Air Transportation Safety and System Stabilization Act, which created the fund in part to protect the airline industry against 9/11 liabilities. Amendments to that act extended those protections to the City of New York, the Port Authority, the Silverstein entities, airport security companies, and even plane manufacturers.

By the time the Virgilio case came before Hellerstein, every plaintiff had either already been compensated, was under review by Feinberg's office, or had signed up but not yet submitted their detailed applications. They had to complete their claims by June, when Feinberg's federally funded operation was scheduled to go out of business. But they and their attorneys were in court, hoping that the judge would allow the lawsuit to proceed anyway. They knew their prospects were bleak. The statute was depressingly clear. Of course, it did not expressly protect Motorola or any company like it, whose "defective product" was accused of contributing to the carnage that day. "Motorola should not be an incidental beneficiary of the fund," said the lawyer for the families. But, as Judge Hellerstein quickly pointed out, "the only exclusion" in the statute "is the terrorists," and "Motorola doesn't fall into that category, no matter what you allege against it."

Hellerstein was sending out all sorts of signals from the bench about how he felt about this troubling case. "The firemen knew for years that the city's radio system did not work very well. It was notorious within the fire department," he said shortly after the hearing began in the late afternoon. "I have to look in the legal sense and the moral sense throughout, because the events of September 11 cry out for justice and equity and morality to the greatest extent possible. And I say these remarks because the audience is made up very much of the people who directly suffered from what happened and I have that in mind."

The city's lawyer led the defense until the judge asked Motorola's counsel, Michael Schissel, if "Motorola had a point of view that it wishes to express."

"No," was Schissel's answer, "I was anticipating my response was the same one" that the city's lawyer had just offered. It was a not-surprising indication once again of the company and city's bonding of interest. Schissel was from the Washington-based law firm of Arnold & Porter, and it had represented Motorola for years. Ironically, it had also represented DOITT on telecommunications issues, getting multiyear contracts that totaled $1.2 million over the Giuliani years.

In fact, Giuliani singled Arnold & Porter out at his 1995 press conference and praised it for winning the Channel 16 waiver from the FCC—the spectrum bonanza that paved the way to the XTS 3500. In 2001, the firm finally stopped its simultaneous representation of the city and companies like Motorola, sensing a possible conflict, according to a DOITT source. Motorola lobbyists in Washington periodically reported that they had acted to support public safety agencies in their efforts to win greater spectrum allocations. But, in this instance, the city had paid one of the company's outside counsels to gain spectrum that would lead to hundreds of millions of dollars in city radio and infrastructure contracts for Motorola. The firm even advised DOITT on its radio and spectrum policies.[9]

Arnold & Porter wasn't the only Motorola face in the crowd. Bob Leonard was there, too. Wherever Sally Regenhard went—and she went just about everywhere 9/11 was discussed—she'd bump into Leonard. He was at other Hellerstein hearings, City Council hearings, public presentations by the National Institute of Standards and Technology, and the 9/11 Commission. Everywhere there was a Giuliani question, Leonard popped up to record the answer. "I went over to him once in Hellerstein's court because I'd seen him so often," said Regenhard, "and said 'are you one of the families?' I knew he wasn't. And he said, 'Oh, no, I work for Motorola.' When I saw him after that I would say, 'Hey, Motorola man.' "

Nothing upset Regenhard more than the internal Motorola marketing strategy document entitled "FDNY Fireground Overview" that was filed as part of the case. Under the heading "Longer Term Direction," the undated document, marked "confidential proprietary," read that "FDNY must move off of the XTS 3500 platform." It noted that the radio had "limited fireground capabilities" and advocated the "cancellation of the XTS 3500 soon after resolution and deployment," suggesting that the company should stop selling the radio once the department finally deployed it. The marketing strategy called for the "next generation FDNY radio—based on XTS 5000," a new digital radio Motorola was developing. It called for shifting the department to the 5000s by the 2004–2008 period with "FULL Fireground features," confirming that the 3500s had "limited" features.

When *Newsday* published the memo, a Motorola spokeswoman said it was written by the leader of a sales team and only meant to foster internal discussion. However, within months of the department fully deploying the 3500s, Motorola discontinued them in early 2004.

The memo's accuracy about future Motorola actions only made its retrospective look at the history of the 3500s even more disturbing to firefighter families. It posed these questions: "Did we sell them [the FDNY] the wrong platform? Did we oversell fireground? Did we ignore these to keep a $10+M order?" It even asked: "Did we misstate TSP benefits?"—a reference to the digital transmission process. That was the kind of tantalizing evidence that convinced the Virgilio plaintiffs that if Hellerstein ever granted discovery, they would find Motorola and FDNY smoking guns. But Hellerstein's rejection of every legal argument the lawyer for the families was making that afternoon convinced Regenhard early in the hearing that there would be no discovery.

When Hellerstein had heard enough from the lawyers, he did something so rare it still fills Sally Regenhard with joy. He invited the families to speak. "I couldn't believe it," says Regenhard. "It was always my dream to have our day in court. He's such a menschie kind of guy. We were so grateful to have Feinberg's gag taken off our mouths."

Eileen and Rosaleen Tallon, the mother and the sister of 26-year-old Sean Patrick Tallon, rose first. A couple of months later, Eileen Tallon would be one of the voices in the audience at the 9/11 Commission hearing who would yell, in her distinctive brogue, during Giuliani's testimony, "Talk about the radios." She'd signed up with Feinberg well before the Hellerstein proceeding that afternoon, but she'd yet to file an application and she wanted the judge to know that she didn't want to ever file one. "We're hoping you would let this lawsuit go through," she said, "and then I would gladly give up, I don't want to be compensated.

"In April 2001, my son came home and he said to myself and my husband and my daughter one evening: Something is going on with the radios. Something very bad is going on with the radios. He said I'm very worried. Even special chiefs and people were getting transferred farther away when they were complaining about the radios. So I said to him: Well you better stay quiet. You need your job. I said I'm sure the people at the top are taking care of the radios. And I have since found out, Judge, that so much wrong has been done about the radios. And that's why I'm here. I'm looking so that people that did wrong with these radios that we could, you know, show future generations that wrong can't be covered up."

Rosaleen Tallon told Hellerstein that they'd signed up with the fund over a year ago. "I never knew this lawsuit was going to happen," she said. "I wish I had. I'm not looking for money. I'm just looking for people to be held accountable."

A retired cop, Leonard Crisci, stood up for his brother John, a former member of the Fire Department hazmat unit. "As you said earlier," he told Hellerstein, "it wasn't proper. The radios didn't work. You knew it. You all knew it. The foot of the blame lies here."

Alexander Santora, the retired former safety chief from the Fire Department, was there for his 23-year-old son, Christopher. His 40 years in the department spoke: "As the chief in charge of R&D for 10 years, all the equipment down to rubber bands were tested. When it came to the radios, back in the late '80s, there was no testing going on. In '93, when the World Trade Center was hit, we knew we had major problems. No testing went on. Just recent[ly] when they introduced, all of a sudden, new radios that didn't work and we almost lost a firefighter in Queens, the radios were pulled back and what was reissued? The same radios. Was there inappropriate behavior? Was there something going on that shouldn't have been going on? When I found out about this lawsuit, I said that's what I want to do." The body of Santora's son had been incorrectly identified after 9/11 as the body of another firefighter from the same company. When the mistake was later discovered, the body was disinterred and buried again. Santora had gone to both funerals.

Then came Captain John Joyce, who'd lost an army he had helped train. He talked about the 3500s that had malfunctioned during tactical training. In full dress uniform on very unofficial business, he said that the city had tried to silence him, to make a deal with him.

"You've heard many stories about 9/11," Sally Regenhard said in the final statement. "The people who are guilty promulgated these stories that, oh, the firefighters could have left, but they didn't. They heard the orders, but they refused. I'm here to uphold the character and the dignity of my son. He was a marine. If he would've heard an order to evacuate, he would have evacuated. Not only that, he loved his life. He never, never would have done anything to commit suicide." Her only evidence that the radio killed her son was that he was gone. Yet she understood that was a far more credible explanation than the one Giuliani routinely offered—that firefighters chose to stay.

When Regenhard was finished, Hellerstein, a Manhattan liberal appointed by Bill Clinton, spoke in anguish. "If you listen to people and you think about what happened," he said, "it's hard to do your job because you're fraught with emotion. We have to do what is right, not only in terms of what we think is right, but also in terms of applying the law that Congress has established. I took this job knowing that sometimes I'd feel happy in following the law, and other times, I wouldn't feel so happy, but I've got to follow the law, and that's what I'll try to do in this lawsuit and I hope I'll do a better job for listening to Ms. Regenhard, Mrs. Tallon, Mr. Crisci, Mr. Santora and Captain Joyce."

He subsequently tossed the case and was affirmed on appeal by a unanimous circuit court.

PART THREE
THE AFTERMATH

CHAPTER 8
GROUND ZERO

WHEN THE PLANE passed by, the rumbling was loud enough to shake the scaffolding. Working on the 38th floor of a building under construction on East 77th Street, Joe Libretti, a 45-year-old ironworker, could tell immediately that there was something wrong. "It just missed the Empire State Building antenna," he recalled. The plane streaked past and, as the construction crews watched in horror, struck the North Tower of the World Trade Center.

Like almost everyone that September morning, Libretti and his friends assumed they had witnessed a bizarre and terrible accident. They were debating whether to go downtown and offer to help, when an American Airlines flight passed them, even lower than the doomed plane that had just crashed. "I mean, we could see people in the window," Libretti said.

By the time the second plane smashed into the South Tower, the men were already on the move. Just as police and firefighters from all over the city instinctively raced to the site of the disaster, the ironworkers knew that whatever was left in Lower Manhattan would require the services of men who knew how to cut their way through huge piles of metal beams. When Libretti got to his union office at 42nd Street, he would find 100 other men there waiting to be told how they could help.

Joe Firth, an official of the carpenter's union, was going through paperwork at the union headquarters on September 11. Almost as soon as the plane hit, the phone began to ring from union members wanting to know what they should do. "A lot of guys headed down to Ground Zero spontaneously," he said, remembering how his men had pushed through the tide of workers and residents fleeing the area. "I know a lot of people were leaving, but I know a lot of construction workers were heading the other way."

By the time the towers collapsed, many of the people standing near the site, watching in horror, were carpenters and ironworkers and teamsters who had come to be of service. Dave O'Neal, 31, who had begun the day cutting structural steel on a project at Carnegie Hall, remembered later that as he stood watching the South Tower fall, the top floors seemed no more substantial than an egg, thrown down from a great height: "Everything just went flat."

IN THE DAYS that followed, Americans tried to make sense of what had happened, through positive action. They raised money, gave blood, and—in thousands of cases—made their way to the site of the disaster and offered to search for survivors. Students at Oakwood College, a traditionally black school in Huntsville, Alabama, watched the towers fall on television, threw shovels and garbage bags into their bus, and drove directly to New York City. "We traveled about 24 hours to get here," said Dr. Gregory Mims, their chaperone. Unable to gain entry at the site, the students—most of whom played in the school band and had had the foresight to pack their instruments—decided instead to show their solidarity with New Yorkers by marching through Manhattan, playing patriotic songs.[1]

Most of the volunteers had to be turned away. The pancaked towers formed an enormous, unsteady pile of twisted metal and concrete rubble, a threat to anyone who tried to maneuver around on it. Police checkpoints blocked access to anyone but emergency workers—mainly firefighters frantically searching for lost comrades. Men like Joe Libretti and Dave O'Neal were also eventually allowed inside because they had the skills to cut through the tons of mangled steel that was impeding the rescue efforts.

Libretti discovered that his 43-year-old brother, Daniel, a member of one of the Fire Department's elite rescue units, was not on vacation, as he had originally believed. September 11 was Danny's first day back on the job. As the ironworkers waited restlessly for the guards at the smoldering attack site to let them in, Libretti's instinct to help became personal. He would search for his missing brother.

O'Neal, a volunteer fireman in his hometown in Pennsylvania, had many friends in the Fire Department, and ten of them were missing. "We tried looking for them but we couldn't find them. We were finding body parts. . . . They were everywhere, everywhere you went." The air smelled like sulfur, he thought. Strange flames were shooting out of the ground—"purples, greens, yellows."

Some estimates would put the number of men and women who took part in the Ground Zero operations in the tens of thousands, including firefighters and police officers, as well as the volunteers who raced in to help during the chaos that immediately followed the attack and the mainly immigrant cleaning crews who would be called in at the end to prepare the commercial buildings and residences for reoccupation. The number included about 1,700 construction workers, and many of them remained at the site for months. They were primarily men who made midlevel pay building the city's skyscrapers, men who kept their families in the faraway but affordable suburbs of upstate New York or Pennsylvania. Many had grown up in the same neighborhoods that filled the ranks of the city police and firefighters and had multiple connections to the departments. Some, like O'Neal, were volunteer firefighters themselves. Some, like Libretti, had close relatives who were among the missing. They would all risk their lives, stumbling through massive heaps of unstable rubble looking for bodies and clearing away evidence of the city's wound. The work would be terrible, full of sights that would reappear in nightmares and waking trauma for years to come.

The word "hero" has been devalued since 9/11. But if all those who put themselves at risk to search for the living, then the dead, count as heroes, the construction workers were the ones who were unsung.

And in many cases, they would become the victims, too.

* * *

WHEN LIBRETTI AND his friends were allowed to enter Ground Zero, a cloud of dust and gases hung over the 16-acre World Trade Center site. Firefighters with dazed expressions wandered about. Fires burning everywhere sent clouds of smoke and foul fumes into the air. The shifting pile of debris, several stories high, threatened to swallow up rescue workers. Libretti hooked up with two firemen searching what was left of the South Tower. They played their lights on a twisted steel beam and discovered that the bodies of three people—two women and a man—had somehow become wrapped around the beam. Libretti and a firefighter tried to free the man by grabbing his arms, but the body fell apart.

"You couldn't walk more than a few feet in some areas without encountering body parts," he said.

As dawn broke, Libretti heard someone call his name, and he turned to see several firefighters from his brother's rescue unit.

"Where's Danny?" he asked. The men just hung their heads.

It was not until dusk on September 12 that Libretti realized he needed to go home, to get the medicine he took for diabetes. He drove two hours to the house in Pennsylvania, where he lived with his wife and three children. He showered, slept for a few hours, grabbed some clothes, and got back in the car. On the morning of September 13, the ironworker was back with the rescuers at Ground Zero. Like O'Neal, he would remain on the job for months, working around the fuming tower of rubble and sometimes sleeping in one of the tents that had been thrown up for workers across the street from the site.

THE MEN—AND a small number of women—who worked at the World Trade Center site were breathing in a toxic stew of smoke, carbon monoxide, pulverized cement, gypsum, PBCs, and other potentially lethal substances. The towers had turned into a mass of metal, dust, gas, fluids, and rubble that contained up to 1,000 tons of asbestos, along with an enormous amount of fiberglass. The estimated 50,000 personal computers the buildings had housed contained at least 200,000 pounds of lead; the light fixtures gave up lethal quantities of mercury; the burning oil from the tower that housed Rudy Giuliani's smashed command center alone contributed more than 150,000 gallons of fuel. The oil fires released large amounts of benzene—exposure to which can lead to leukemia—while the burning plastic from furniture and carpeting and cable sent up plumes of smoke that laced the air with cancer-causing substances such as dioxins.[2]

It was hard to think about any of that while there was the possibility that people might be trapped beneath the smoldering rubble, praying to be rescued. The emergency services workers had always embraced an ethic of disregarding personal danger in order to save other people, and the construction workers on the site followed their lead.

On that first day, 17 people were pulled from the rubble—14 of them from one spot in the North Tower where a section of stairwell had remained intact. Everyone imagined that was just the beginning. When Pasquale Buzzelli, a Port Authority employee who had miraculously survived the collapse of the building with some cuts and a broken foot, was rescued and brought to the hospital emergency room several hours after the tower fell, the staff patched him up quickly and sent him home—certain that all their attention would soon be required by victims who were in far worse shape.[3] But Buzzelli would turn out to be the second-to-last person to be taken out of the building alive. Genelle Guzman-McMillan, another Port Authority worker, was found on the second day, lying next to a dead fireman, her legs crushed and pinned under the debris. Hers would be the last desperate pounding, the last call for help, and the last good news to emerge from the pile.

Everyone who remained unaccounted for was dead. Neither the rescuers nor the relatives of those who were lost could accept that reality. Dr. William Trolan, a physician who came to New York as part of a team of veteran rescuers from California, was haunted by the knots of people who would stand behind rope lines set up by the police, pressing pictures of their relatives on the Ground Zero workers and crying, "Did you see him? Did you see him?" Some were crying. "You feel totally helpless," he said. "You don't want to tell these people there's no hope."

Harold Schapelhouman, who headed the California team, was struck by the pulverizing effect of the height of the towers, and the force of their collapse. "You didn't even find pieces of broken glass," he said. Schapelhouman had participated in rescue operations after the Oklahoma City bombing, where his team had found, if not living victims, at least intact bodies. In New York, he recalled, "we went through decontamination one night and one guy had a gold crown stuck in the sole of his shoe. I mean, that's the degree of what we found."

Of all the could-have and should-have analyses of what happened in Lower Manhattan after the World Trade Center towers fell, very few critics have faulted the city for failing to make the rescuers observe proper health and safety procedures on those first desperate days. No one would have listened, and if the rescuers had been reminded that they were risking their lives, they would have kept going anyway. No matter what protocols are put in place for homeland security, if another terrorist attack happens on American soil, people will rush in to try to save the victims, ignoring their own well-being and leaving officials to simply try to keep as much order as possible at the scene.

But as time elapses and rational hope fades, caution and care might be expected to replace panic and passion. Bruce Lippy, an industrial hygienist who studied health conditions at the site, said international protocols acknowledged that the chance of rescue exists for up to two weeks. At Ground Zero, the chances dimmed far quicker than that. "We expected transition to occur in an ordered, organized, planned and effective manner," said Lippy. "It did not occur for several months."

The only man who could have forced it to occur earlier was Rudy Giuliani. Not only did he alone have the legal authority to shift the operation from rescue to cleanup and recovery, his instant icon status and symbiotic identification with the firefighters gave him the moral authority to make it happen. Only he could have effectively put the health of the thousands at the pile ahead of the hunt for the remains of those already lost. Phil McArdle, the health and safety officer of the firefighters union, would later assess the long-term health effects of the months at Ground Zero, likening it to Agent Orange and Vietnam. "We've done a good job of taking care of the dead," he said, "but such a terrible job of taking care of the living."[4]

The mayor certainly knew about the dangers the polluted air posed to workers at the site. After the very first Giuliani cabinet meeting on the attack, Fire Commissioner Tom Von Essen noted in his diary that "asbestos will be a problem" and that the men working at the core would need respirators. Von Essen soon got the "World Trade Center cough" himself, even though he only visited the site periodically and spent most of his day "hacking," as he put it, at Giuliani's side at the Pier 92 command center. While Von Essen's cough went away after a few weeks, he said he was sure that "some" at the site would have "long-term lung problems."

The mayor was also under no illusions about the chances of finding anyone alive after the first few days. "A collapse expert from FEMA had warned at an early staff meeting that it was a virtual certainty that no one had survived the titanic force of 1,300 feet compressing into a mere 80," wrote Von Essen, acknowledging that they knew "the rescue was over." The firefighters did not acknowledge that reality and would resist any change as a betrayal of their mission, but Von Essen understood that it had to happen and pushed Giuliani to do it again and again.

There should have been a point—perhaps as early as the end of the first week—at which Ground Zero was officially reclassified as a demolition, debris removal, recovery, and reconstruction site involving hazardous materials.[5] The workers needed to be fitted for the proper equipment, told that they had to wear it at all times, and encouraged to pace their efforts to accommodate respirators and other bulky gear. They needed to be given regular medical checkups and forced to wash the dust off their bodies and change clothes before they went home and contaminated their families.

Von Essen wanted to admit publicly that the rescue operation was over. He planned to break the news to the families of the missing firefighters at a meeting on September 18. But as he was on the way to the podium, Giuliani grabbed his arm and whispered, "Don't say it.

They aren't ready." The fire commissioner noted in his diary on October 14—after particularly troubling incidents at the site—"mayor now seems ready for what I recommended two weeks ago . . . now we need to figure out how to do it." In the end, there was never any official announcement. Instead, the operation continued 24/7—half cleanup, half-body-part-search, governed by its own rules rather than those required for a hazmat event. In late November, Giuliani got into a sharp exchange with a construction worker who called the mayor's radio program to ask that the crews, who were working through Thanksgiving, be given Christmas day off to spend with their families. The mayor refused, because, he said, the families of the police and firefighters who died at the site "feel so strongly" that the recovery of the bodies had to keep going on around the clock.[6] When city officials on site closed it for Veterans Day earlier that month, Giuliani deputies made it clear that they should not shut it down again.

A WEEK AFTER the attack, the U.S. Environmental Protection Agency (EPA)—in a press release geared to reassure the public that Lower Manhattan's air was safe—said that the highest levels of asbestos "have been detected within one-half block of ground zero where rescuers have been provided with protective equipment." The notice drew a picture of a well-ordered area, with everyone assigned the level of precaution appropriate to his or her situation. That was far from reality. The air quality in residential and commercial neighborhoods that were being reopened was a mixed bag at best, and not all—or even most—of the people at Ground Zero were wearing protection. The failure to require the proper health and safety gear at the site was inexcusable, wrote Dr. Philip Landrigan of Mount Sinai Hospital, and "the result will almost certainly be unnecessary disease and death."[7]

Obviously, if the Giuliani administration had truly been worrying about and preparing for a terrorist attack in the years since the first World Trade Center bombing, the Office of Emergency Management, the Fire Department, or other city departments would have had more and better equipment for this kind of emergency. But it hadn't, and the only protection Libretti was offered in the early days at the site was a small dust mask that covered the nose and mouth. The masks were not intended for the kind of air the workers were breathing at Ground Zero, and they quickly got filthy and clogged by debris. Yet, according to the three doctors who ran the FDNY's Health Services Bureau, 70 percent of the firefighters "had access to only a dust mask not approved for this type of exposure" on September 11. The same doctors said 82 percent of firefighters over the course of the early days were "without respiratory protection," some eschewing the inadequate masks they were given.[8]

Phil McArdle and Uniformed Firefighters Association president Steve Cassidy wrote the Fire Department after 9/11 complaining that "for years" the department had "ignored many issues related to respiratory protection" and that "these lapses in responsibility" had a "detrimental effect on firefighters" at Ground Zero. The only type of protection addressed in the circulars sent to all units, the union letter contended, was the mask normally worn at fires—a bulky "self-contained breathing apparatus" with an oxygen tank that typically gave firefighters about 18 minutes of fresh air. The kind of long-term, air-purifying respirators needed at the site were rarely seen at the FDNY, the union leaders said. Most members hadn't been "trained properly" in how to use them, nor had they been fitted for them—and if a respirator did not fit, it could be worse than useless. Better protective gear for protracted hazardous exposure had been one of the post-'93 bombing recommendations—generated by then chief of department, Anthony Fusco—but there had been little practical change in eight years.

The National Institute for Occupational Health and Safety had 15 hygienists at the site on September 13, sampling the air and trying to determine the level of danger. Almost immediately everyone agreed that the workers should wear dual-cartridge air-purifying respirators that provided extended protection, covering their noses and the bottom half of their faces. The federal health experts from the Occupational Safety and Health Administration (OSHA) and other agencies threw themselves into the project, and the site was soon flooded with tens of thousands of these respirators, many arriving by the end of the first week. But some fit badly, and the men disliked putting them on. "The smoke was so intense that once you're in that air, your mask would fill up with smoke. You'd have to take your mask off or otherwise you couldn't breathe," said Dave O'Neal.

McArdle and Cassidy, the firefighters union leaders, conceded that most of the respirators "went unused during the weeks of the initial operation," and they blamed that on the failure of "supervisors and safety officers." Von Essen's diary for October 14 said almost the same thing after recounting incidents at the site: "Another glaring example of lack of discipline and leadership by some officers in command at the site . . . proper protective clothing, training, position, you name it." The Fire Department doctors later wrote that only 30 percent of the firefighters "were able to wear the respirators most of the time," describing them as "impossible to wear during prolonged work activities" because firefighters couldn't talk to each other or exert themselves.

Harold Schapelhouman's rescue team from California had been working at the site for days before they were instructed to insert a second cartridge into their respirators that would filter out dangerous chemicals.

"Usually you err on the side of safety and then you may decrease," he said. His men, he recalled, asked each other, "What? How could this be?"

The workers who naturally preferred to believe they could work unencumbered got lots of reinforcement. "Supervisory personnel from many organizations on site regularly entered the restricted zone without respiratory protection, setting a terrible example for their workers," wrote Bruce Lippy. "Political figures, dignitaries, and movie stars visiting the site had to be persistently hounded by safety professionals to even wear hardhats. Miss America participated in a photo opportunity inside the restricted zone, reportedly in heels. . . . Poor role models abounded."[9] The most important role model of all, of course, was Rudy Giuliani, who brought the biggest dignitaries on Ground Zero tours himself. And the mayor, of course, was never seen on television or in news photos wearing a respirator (although he did wear a face mask on occasion).

After the first few days, most people who walked around the periphery of the pile were breathing air that tests generally showed to be fairly safe. The greatest peril lay in the "green zone" where fires continued to burn below the surface, consuming plastics and oil and god knows what, and where workers continued to dig deeper into the ruins, raising new levels of dust. But all attempts to define exactly where the green zone began came to naught, and the plume of smoke from the pile continually shifted with the wind, contaminating the air in different sections of the work site. The difficulty in making sweeping judgments about which parts of the site were safe was illuminated when Juan Gonzalez reported in the *Daily News* that on some occasions in October, the EPA had measured levels of benzene—a known carcinogen—from 16 to 42 times higher than the permissible federal limit. An EPA spokeswoman agreed the readings were "high" in the area right around the still-burning fires, but noted that they dropped dramatically outside the plumes of smoke. Then she acknowledged that the winds sometimes brought the plumes right in the vicinity of the workers.[10]

In the midst of all this conflicting information, the workers seemed to hear only the reports that suggested they could continue the jobs they were bent on completing without the uncomfortable respirators. Hygienist Lippy said the construction workers he observed were so determined to be upbeat about the air that they frequently let their respirators dangle from their necks "like loose neckties," with no supervisors taking notice. There were certainly plenty of reassuring reports to hear. "I'm not saying the EPA lied to us—I don't know that," said Schapelhouman. "I don't know what the truth is. All I know is one thing: 70 percent of my people were sick when they came back, a number have continued to be sick over the years and there's a big question mark in my mind."

But it wasn't just the EPA. The city health commissioner, Dr. Neal Cohen, said there were "no significant adverse health risks." Joel Miele, the environmental commissioner, said "It's not a health concern." And Giuliani himself argued that the stench and air were "not health-threatening," encouraging people to ignore what their own senses were telling them. "It may be uncomfortable and it may be offensive—and it is in many ways," he told CNN, "but the reality is it's not dangerous."

It would have been an important signal if Giuliani, who was constantly surrounded by Von Essen, Police Commissioner Kerik, and OEM's Sheirer at his daily televised briefings, had occasionally featured health experts to address the air quality issues and carefully differentiate between the pile, the neighborhood, and nearby areas. The Natural Resources Defense Council, a leading environmental group, called Giuliani's frequent press statements "inspirational, comforting and universally welcomed," but said they "fell short of the mark" when it came to communicating about health matters. The NRDC said the city "could have called upon independent medical experts based at the most prestigious hospitals and universities to help explain available data." Instead, everyone from Henry Kissinger to Jesse Jackson showed up at Giuliani's side at the press briefings, and the city failed to reach out to these experts.

It would have been just as important a signal for Giuliani to don a respirator when he actually went to the site, to show those closest to the pile that they should not feel they were being weak or malingerers if they wore the proper protection. But the mayor was interested in sending a different signal altogether: that Lower Manhattan was open for business and safe for both Wall Street workers and residents. That would certainly not have been reinforced by the sight of a huge contingent of workers clomping around Ground Zero in protective gear and respirators. (When EPA workers erected a huge tent where workers could wash off their boots and shower after a shift, one of them remembers being instructed not to call the area a "decontamination" station.)[11] A volunteer who worked on the site and developed long-term respiratory problems remembered hearing the mayor assure the public that the air downtown was safe. "When you have someone of the caliber of Mayor Giuliani saying it, they took that as gospel," he said.

About a week after the attack, Giuliani's health commissioner, Cohen, told reporters that the city would put more safety officials at the site to make sure searchers wore the proper equipment. But as late as October 5—24 days after the attack—EPA chief Bruce Sprague sent a letter to a Cohen deputy deriding the city's "very inconsistent compliance with our recommendations," listing the lack of respiratory protection and personnel wash stations. Sprague wrote the letter after raising the subject of respirator noncompliance in a couple of conversations with a city health official. "It's almost common sense," Sprague said in a subsequent court deposition, "that one does not stick their head over a barbecue grill for hours and hours and then expect there not to be some sort of an issue." Though Sprague pushed the health commissioner to issue an order requiring respirator use, it wasn't done. (When EPA Administrator Christie Whitman saw Ground Zero footage on CNN late one night, she was alarmed by the sight of workers laboring without respirators, and started making midnight calls to emergency officials to try to shake things up.[12]) Since the EPA did not have the authority to enforce safety policies for anyone except its own employees, Sprague urged the city to have the incident commander adopt and enforce a safety plan. The commander at the time was the highest-ranking fire chief on site. Later in October, the top officials of the city's Department of Design and Construction (DDC) were named co-incident commander with the fire department. But Giuliani himself was always regarded as the incident commander by the federal agencies that deferred to him. And Mike Burton, the executive deputy commissioner of DDC who actually supervised the demolition and cleanup, testified later that "ultimately the mayor gets the responsibility for what happened down there." Burton said "everything was coordinated" through his boss Holden, "deputy mayors and the mayor."

Though the construction giant Bechtel was hired by the city as its health and safety consultant, no one paid attention to its late September initial recommendation that the number of entry/exit points onto the site be drastically reduced from 20 or so to 2. That was Bechtel's way of forcing workers to be fitted for respirators when they arrived and decontaminated when they left. Incredibly, Bechtel didn't announce a comprehensive health and safety plan until October 29.

Soon after, it left the site altogether, replaced by another vendor with another approach.

NEW YORK CITY is in many ways a nation-state. It is so vast and—by American standards—so old, that it has, over time, come to assume responsibilities for many functions that other cities have never thought of doing for themselves. "None of us wondered, 'Should we contact the state? Should we contact the feds? FEMA? The Army Corps?' It was just 'We've got a disaster here. Let's fix it.' It was instinctive," said Ken Holden, the commissioner at DDC, which put itself in charge of clearing the site. The DDC did an excellent job with many of the responsibilities it assumed, but Holden regarded the respirator issue as largely beyond his control, particularly in the case of firefighters and police officers who would only take orders from their own brass.

Ground Zero had been divided into four segments, each assigned to a different construction company. Holden's staff gave the contractors guidelines about where the workers should take precautions against possibly toxic air, and the contractors were expected to follow through. But the incentives went in the other direction. The contractors were being pressed to complete the work quickly and were promised by the city that they would be indemnified against any legal claims from workers who felt their health had been harmed at the site. While the extraordinary crisis may have justified this extraordinary exemption, it also meant that the contractors had no financial incentive—as they ordinarily do at a job site—to protect workers. So, on October 15, even as the city was publicly rejecting any suggestion of a health threat at the site, its top lawyers prepared a letter recommending indemnification for the contractors. Signed by Holden, the letter said that the city and the contractors would "be open to exposure years after the Project is completed based on hazardous materials claims." It predicted that the claims could be so substantial they could "bankrupt" the city or the companies.

The Giuliani administration was also aggressively pursuing federal legislation to limit the city's liability. As early as October 4, City

Comptroller Alan Hevesi had issued a report on the impact of 9/11 on the city's budget and cited its potential liability for workers' respiratory illnesses. Even as every public pronouncement from the mayor and his men was filled with reassurance, an internal memo to Deputy Mayor Robert Harding reported that the Law Department was estimating that there were 35,000 potential 9/11 plaintiffs against the city. Two of the prime potential claims cited in the memo—just weeks into the cleanup—were that rescue workers had been "provided with faulty equipment or no equipment (i.e., respirators)" and that Ground Zero was an "unsafe workplace" under various federal safety and labor laws. The memo urged the city to push Congress to create a fund to cover the city's liability similar to the 9/11 Victim Compensation Fund it had just created for the airline industry.[13] On November 1, Giuliani wrote a letter to Congress supporting an amendment to the Victim Compensation Fund legislation that would extend its benefits to those working at the site, as well as fix a $350 million cap on city liability. In late November, an amendment passed achieving both purposes. Those who could make a provable case of injury at the site were added to the list of eligible fund recipients, if they, like the families of the dead, surrendered their right to sue.

For all the tragedy the terrorists had created in Lower Manhattan, everyone involved with the recovery was aware that things could have been far worse. There were no "dirty" bombs on the planes and therefore no radiation. The huge tanks of Freon stored below the towers for air-conditioning did not explode. And the so-called "bathtub"—a massive concrete wall that surrounded the World Trade Center's seven-story basement and prevented the complex from being inundated by the Hudson River—did not rupture. After the towers collapsed, support for some parts of the wall was provided only by the tons of debris that had fallen into the basement. Careless removal of the debris could have caused part of the wall to collapse inward and, in the worst-case scenario, send in torrents of gushing water that would have drowned the workers in the basement, flooding the railroad tunnels, the subway system, and even part of the city.

Strangely, the mayor paid no attention to the bathtub crisis, according to Holden, who appeared twice a day at the command center for cabinet meetings early on, but was eventually excused from all but occasional meetings. "Daily meetings were held with Commissioner Kerik and Von Essen, not really with me," he said. The mayor never spoke to him individually about any site issues, nor was he even told about the EPA letter to the health department that complained about the lack of wash stations and protective gear.

"We would have loved to have had the authority to throw out people who would not wear protective equipment," recalled Holden, adding that at one point he told his top aide, "If rates of use don't improve, I will shut the site completely." Asked if either Giuliani or the deputy mayor who oversaw his agency ever "showed any interest in the exposure of the workers" to Ground Zero health hazards, Holden said: "Not that I recall. Not specifically concerning the wearing of respirators." Holden pressed, like Von Essen, to reduce the number of firefighters on the site, but Giuliani rebuffed his efforts as well. Instead, the mayor was "mostly interested in reports on the tons of debris and steel that were shipped out and the bodies that were found," says Holden. Giuliani wanted "numbers and results, not what type of work, problems and risks were behind those numbers." Holden's deputy Burton knew Von Essen very well and sent digital photos of firefighters without respirators to him in an effort, as Burton later testified, "to try to up the utilization rate." But Burton's memos, phone conversations and photos provoked no discernible Fire Department action.

The mayor's single-minded focus was on getting the cleanup done. He told the city that Ground Zero would be cleared in 180 days, a deadline in search of a reason. The sense of rush meant that even after any chance of rescue was over, there was not so much as a discussion about reorganizing the cleanup around sounder health and safety standards. "Giuliani wanted us to clean it up quickly," said Holden. "The sooner the better. He did not want to leave a visible sign of the city's open wound." But it was not as if the mayor was the only one who was hell-bent to get the job done, and Holden said he did not feel he was being directly pressured to hurry. "It was the spirit of what was happening," he said. "We started at 24/7 when there was an emergency, and we kept on going after that."

In retrospect, the rush—and the lack of concern for the workers' health and safety that went with it—seems manic, an emotional stampede. The people involved at the time may have felt they needed to close what Holden called an open wound. But five years later, the most contaminated office tower near the site, the Deutsche Bank building, was still awaiting demolition and Ground Zero remained a vacant construction site, the planned Freedom Tower still a mirage. A brief hiatus, or the rapid introduction of the rules suggested by Bechtel, or any form of respirator enforcement with penalties, could have made all the difference. It would have temporarily slowed the firefighter-driven search for body parts, but spared the health of so many of the searchers. Adjusting the search and the cleanup to the reality of the fire—which kept burning and spewing toxins until late December—would have saved lives, and prevented disabling illness. Alternative ways of putting out what became the longest-burning commercial fire in history, such as foam and nitrogen, could have ended the toxic smoke sooner. But they were rejected—in part due to concerns about disturbing the buried human remains.

The mayor did express interest in construction safety issues. He told Holden that he did not want one more death at the site and, remarkably, no one was killed during the entire rescue, recovery, and cleanup effort, although some of the workers did suffer serious injuries. That was no small feat. The site was unpredictably dangerous. Lippy, whose union members operated the very large machines atop the rubble, summed up the hazards with one worker's alarming saga—he'd gone off for coffee only to discover, when he returned, that his vehicle had literally been swallowed up by the pile.

THE DAYS BLED into one another. Workers came and workers went, but some, like Libretti and O'Neal, found it impossible to leave. Every body part discovered, they felt, meant a possible sense of closure for a grieving family somewhere. Every ton of rubble carted away was a little act of defiance toward the terrorists. One of the ironworkers,

Libretti remembered, hurt his leg but refused to leave the site to get it checked. He continued working for a week before they discovered it was broken.

Years after the disaster, Libretti was still haunted by the things he had seen at Ground Zero. A sight or a smell, like a rotting piece of roadkill on the highway, would trigger a flashback. And in the dark, there were the nightmares. "I'll go to sleep and I'll toss and turn until I do fall asleep and then I'll wake up because I'll smell rotten flesh. It's like I'm right there. Or I'll just get a flash of when we picked a guy out of the elevator—he was burnt to a crisp. Or the mush that I fell in, which happened to be a pile of skin. And every time I don't shave and let my beard grow right here, right where my chin was in the mush, I now get a rash."

In the wake of the disaster, there were many programs set up to help firefighters, police, residents of Lower Manhattan, schoolchildren, or anyone else traumatized that day to recover their mental equilibrium. There was less discussion of the mental health of workers charged with cleaning up the site. A screening program set up to evaluate the health of rescue and recovery workers found long-term rates of post-traumatic stress, panic, and anxiety that were much higher than in the general population. "Construction workers are a particularly vulnerable population," said Lippy. "They don't talk about what they've seen, they want to tough it out, which is the wrong way to go with the kind of horrors they experienced."

And as the weeks and months wore on, they also didn't talk much about how hard it was to breathe. Neither the work culture of the site, which had been established in those early desperate days, nor the experience of the workers—who generally worked on construction projects that offered no air hazards—encouraged the men to protect themselves. Even when federal agencies finally assumed a more active role, the intervention was too mild to get their attention. As much as city officials later tried to blame the Occupational Safety and Health Administration (OSHA) for the lack of respirator use, they acknowledged in court testimony that OSHA agreed in September to act merely as an adviser. Finally, on November 20, 2001, OSHA signed an agreement with the city and the contractors, explicitly yielding its power to issue fines for health and safety violations in return for a promise from the contractors to respond to requests from OSHA compliance officers. This memo of understanding obviously left those compliance officers rather toothless when it came to enforcing their concerns. There is some evidence the deal actually worked on some levels, such as limiting accidents. But it did not improve the shocking infrequency of respirator use.

Once the debris had been taken off the site—an operation that always entailed raising dust from new layers of as-yet unexplored wreckage—it was carted off to the Fresh Kills Landfill in Staten Island, where everything was carefully examined for additional body parts. Lippy observed that although the debris was wetted down to protect the workers at the second site from dust, almost all of them wore respirators. "Why could we get 90 percent respirator compliance at Fresh Kills and only 30 percent at Ground Zero?" he asked.[14]

But the same disconnect was apparent at the site itself. Dave O'Neal remembers watching the safety monitors and thinking they looked like beekeepers in "those white outfits, with tanks all hooked up to them." They were walking around like the actors in a bioterrorism movie, covered head to toe. The rescue workers, meanwhile, were wearing their regular clothes, with no protection at all: "We're taking our shirts off, we're sweating our asses off. We're trying to get these [body] parts out to get these people closure at home."

When O'Neal pointed all this out to a firefighter friend who was working on the site, the officer said fatalistically that the people who said the area was safe had undoubtedly been wrong or lying. "But what are we gonna do? We've already been exposed. You know there's nothing you can do to change it."

On September 27, O'Neal says, he coughed up some blood and was sent to an on-site medical facility. "It was like a little hospital inside, but it was a trailer. And this guy came up to me and said you just got a little irritation in your throat, it's just from the dust. You'll be fine. Wear this mask, take this medicine. It was Robitussin."

One night, when Libretti and some other workers were sitting by a crane eating chicken and Jell-O in the middle of the myriad health hazards triggered by the destruction of towers full of asbestos, computers, oil, and other environmental gremlins, a supervisor came up and shined a flashlight at them. "What are you doing?" he demanded.

"We're having dinner," they told him.

"You can't eat out here," said the horrified newcomer.

"We've been doing it for a month," one of the men said.

VON ESSEN WAS concerned by what he saw at the site. "Though many workers had come down with the World Trade Center cough, many still refused to wear their masks and respirators," he wrote in his autobiography. "People walked around without hard hats and safety glasses." And he, too, noted the way workers were casually eating at the pile "without even washing their hands." But any attempt to enforce more discipline would have had to include a reduction in the number of firefighters and police who were on the site, something Giuliani resisted.

The Fire Department, Police Department, and Port Authority Police were each permitted to have 75 people on the site at all times, looking for remains of the victims. It was, to the participants, a holy mission. But the Ground Zero workers—particularly the firefighters—went at it with a kind of zealous abandon. They often refused to wear any kind of protective gear that would slow down their search, putting themselves in peril. On October 31, Giuliani finally announced that the number of workers would be scaled back to a third of the former level. As everyone had feared, the change precipitated an emotional, somewhat violent protest from the firefighters. On November 2, several hundred off-duty firefighters marched on Ground Zero to demand that the search for bodies continue as before. "Bring our brothers home!" they cried. When the police attempted to stop them from entering the site, fighting broke out and five officers were injured. Twelve protesters were arrested. The demonstrators marched on to City Hall, chanting "Rudy must go!"

The mayor's wrath was first directed not at the firefighters but at a group of ironworkers who were filmed mugging for the cameras and waving an American flag. Giuliani called Ken Holden and demanded that he fire the offending ironworkers. Holden, who recalls it as the only time Giuliani ever called him during the cleanup, managed to avoid taking any action, heading off what would certainly have been a dangerous collision between the mayor and the proud, exhausted construction workers. Eventually, the fire unions apologized to the police and the mayor dropped the charges against all but one of the demonstrators and returned the size of the firefighters' search teams to previous levels.[15]

The mayor insisted he had made the important point. "This is all about safety," he said. "We don't want any more casualties. We don't want any serious injuries. This is a very dangerous operation. We have to make sure everybody's wearing equipment, everybody's being careful, and everything's being done in a coordinated way."

For many of the workers, it was too little, too late. It also didn't last too long. Within nine days, the mayor approved more firefighters at the site than had been there when the Halloween plan was announced.

It was around the same time that Libretti remembers finally being fitted for a respirator. "And from that point on everybody had to wear masks on the site. But by that point, everyone in the crew I had worked with had gotten sick. One after another. Got the runs, coughing. It just ran through us. Everybody had it."

In March, about half a year after the towers fell, the pile of rubble had become much more like a normal construction site. Dave O'Neal was handed "a regular plain piece of paper, and it said well, if you ain't wearing your mask, it's a $3,000 fine. If you're not wearing your goggles, it's a $1,000 fine." He had been working at the site for months. The man who gave him the notice, O'Neal said, told him that 7 World Trade Center, where he had worked, "had transformers that were burning that is totally toxic to your health, that can give you cancer and everything else. And I says: 'Well, why didn't they tell us that when we were there? Maybe that's the reason why I was getting dizzy. Ninety percent of the time I was getting migraines.' And they said: 'Well, then we didn't know.' "

On March 5, 2002, O'Neal was working a midnight-to-noon shift, burning steel, "and I was coughing a lot, real hard, like I had laryngitis and I couldn't speak and I was getting pounding migraines." A friend brought him some Tylenol from the medical facility for the headache, and O'Neal suddenly felt nauseous.

He moved away from the work area and vomited up blood.

"Something's definitely wrong," he told himself.

Covered with blood and weak-limbed, O'Neal suddenly had a vision of himself collapsing and being mistaken for one of the World Trade Center corpses. He asked a friend to walk him to the medical facility, to make sure that if he fainted, someone would revive him and carry him to safety. As soon as he walked into the makeshift clinic, he was dispatched to a hospital emergency room.

As O'Neal remembers it, after he was examined, a doctor came and said, "I have bad news."

"What do you mean, bad news?" O'Neal asked.

"You have some kind of lung disease," the doctor said. "How old are you?"

O'Neal was 31.

"Both of your lungs are black," the doctor said. "You'll have to have surgery."

The doctor's prediction turned out to be more than right. O'Neal wound up undergoing multiple surgeries and enduring bouts of pneumonia that left him totally disabled. He suffers from a nervous condition that leaves him shaking as if he had Parkinson's disease. His younger daughter, who was a baby at the time of the World Trade Center attack, also suffers from chronic illness, and O'Neal worried that it might have come from dust on the clothes he wore when he returned from the site. His wife, who has back problems as the result of an auto accident, cannot work. The family lives on workers' compensation and Social Security disability payments, but neither his wife nor his daughters have health coverage.

At the time of the attack, O'Neal was a lean, 160-pound young man, but the medications, which include steroids, and the lack of activity due to his lung problems caused him to balloon to nearly 300 pounds. The men he keeps in touch with from his days at Ground

Zero also report a wide range of problems. "My friend Billy, he's got lung disease. Joe, he's got lung disease and 20 other health issues. His teeth are falling out from being down there." And O'Neal still has nightmares. "I'll wake up in the middle of the night sweating. I'll hear voices. I'll hear cranes moving. The sounds of metal and screeching."

THE CONSTRUCTION WORKERS who had labored at Ground Zero had every reason to be proud of what they had achieved. Although the city's real estate interests had always complained about the trouble they had handling the unionized workers on their projects, the teamsters, ironworkers, carpenters, and other skilled workers at Ground Zero exhibited a dedication and work ethic as intense as those of the fire-fighters. Under conditions of great hardship and real danger, they had pulled 1.5 million tons of steel, concrete, and other rubble from the site of the disaster. The work did not end until May 30, 2002, when Mayor Michael Bloomberg presided over a modest ceremony, watching the last load of steel from the South Tower being removed from what by then looked in many ways like an enormous, but otherwise unexceptional, construction site.

Observing his fellow members of the carpenters union, Joe Firth felt that men who had been held together by the shared sense of mission at the site tended to fall apart once their services were no longer needed. "A lot of guys, after it was all over, they felt like they'd lost a part of themselves. I think that's when the reality set in of what we'd gone through," he said. For some time after he left Ground Zero, Firth found himself breaking down whenever he drove across the George Washington Bridge. "I remember coming across and seeing the smoke and I'd get flashbacks . . . of the first body I'd found," he said.

Ground Zero follows everyone who worked there. "They tell you in EMT school never to look into the eyes of somebody who's dead because they kind of haunt you. On that particular day it was a lot of eyes looking at you, no matter which way you turned," said John Graham, an emergency medical technician.[16] But the burden lies far more heavily on some than on others, both mentally and physically. Some combination of luck and genes causes respiratory disease to pass over one individual and come crashing down on another. "I've seen two guys work side by side and get the same exposure and 15 years later, one's got crippling scarring in his lungs and the other's barely affected," said Dr. Stephen Levin of Mount Sinai Medical Center, an expert on the effects of asbestos on construction workers.[17] Levin told a congressional subcommittee that Mount Sinai "saw people being taken off the pile within the first couple of days, gasping for breath, choking, and could predict at that time that there would be a great deal of potential long-term effects with respiratory problems."

Three years after the terrorist attack, Mount Sinai released a report on a screening program involving more than 1,000 people who had worked or volunteered at Ground Zero. Nearly half had "new and persistent respiratory problems," such as coughing, shortness of breath, hoarseness, ear pain, or nosebleeds, said the Centers for Disease Control, which funded the program. More than half had persistent psychological problems, including panic attacks and flashbacks. The percentages roughly held when the clinic had finished examining nearly 15,000 workers. In the final six months of 2005, Mount Sinai treated 841 Ground Zero veterans, and an astounding 67 percent of them were laborers or construction workers, with 85 percent having "upper airway illnesses."

About 10,000 people who worked in or around the site have applied for workers' compensation, claiming they had been left unable to work. That did not include the firefighters, about 320 of whom retired because of breathing problems. At the end of 2005, more than 400 others were on restricted light duty or not working at all because of 9/11-related health problems. Dr. David Prezant, codirector of the Fire Department's World Trade Center medical program, tested 13,000 firefighters, emergency medical technicians, and paramedics and found that after 9/11 their average breathing capacity dropped more than 12 times what it would have during the normal aging process.

Harold Schapelhouman's rescue team split up when it returned to California, the men returning to their jobs in 16 different agencies around the state. It was not until they held a reunion barbecue that he realized something was wrong. Although the turnout was good, a lot of the absentees had been unable to come because of illness. "We had a percentage in there of people who were sick—and not just a little sick. A number of those guys had pneumonia. They were younger, very strong, viable guys who normally you wouldn't anticipate that would happen to." One of them, Dr. Trolan, has had three cases of pneumonia since 9/11—"always in the same spot." In 2005, his physical showed he had restrictive airway disease that had been growing progressively worse over the last three years.

The starkest evidence of the damage done, however, comes from a source close to the mayor. Ken Feinberg, the special master who ran the Victim Compensation Fund, has been a Giuliani friend since the '70s and was selected for the post in part on his recommendation. Feinberg awarded over a billion dollars in compensation to 2,680 claimants with physical injuries, after painstaking reviews of their individual applications. Uniformed fire, police, Port Authority, and emergency medical service workers collected 70 percent of the total, with 1,388 firefighters receiving $626 million. Only 103 construction workers won compensation, tallying $32 million. Unsurprisingly, nearly 55 percent of the award was for respiratory injury, with another 27 percent for multiple injuries, including respiratory. The average award was for nearly $400,000. While workers were supposed to seek medical treatment within 72 hours of the attack to qualify for an award, Feinberg was empowered under the regulations to grant waivers. His final report indicated that he'd done that for "hundreds of rescue workers who were diagnosed with demonstrable and documented respiratory injuries directly related to their rescue service."[18]

A city attorney says the Feinberg awards reduced the number of Ground Zero claimants against the city by over 1,200. Nevertheless, by 2006, 492 rescue workers, including many construction workers, were still suing.[19] The city filed a motion before U.S. District Court Judge Alvin Hellerstein to dismiss the cases, and more than four years after the attack, both sides were still submitting papers. In an earlier attempt to sue federal EPA and OSHA officials, which Hellerstein dismissed, the judge praised the people who risked their lives and health at the site, but added that the fact that some of them wound up suffering terribly "doesn't mean there's a remedy." David Worby, one of the lawyers in the city cases, has brought a class action suit on behalf of 7,300 people. Worby claims that 23 to 40 people have already died as a direct result of their labor at Ground Zero, and the media has begun to take note of the stories his clients stepped forward to tell about their compromised health, their ruined careers, and their lost husbands, sons, and fathers.

Joe Libretti is sitting in a house in Pennsylvania, fighting chronic lung disease and depression. He worries about his family, medical expenses, and hanging onto his home. "I can't run up the stairs. I can't carry anything heavy," he said. Always a physically active man, he gained weight from inactivity. He did not qualify for workers' compensation benefits until the end of 2005. "I've used up money I put away. I used up my annuity. My wife worries about everything," he said.

No one intended that any of the workers who spent months in one of the most physically unpleasant, emotionally traumatic, and potentially dangerous sites the country has ever seen, striving to recover bodies and restore a sense of safety and normality, would be left to suffer from the health consequences without adequate compensation. Yet, for many, that seems to be what is happening. Of the more than 8,000 people who filed claims for workers' compensation with the state of New York, citing injury or exposure to toxic air from Ground Zero, nearly a third were challenged. The workers' compensation system and the federal Feinberg program are based on the presumption that injuries at Ground Zero would become quickly apparent. But often, they didn't. Some workers were later able to get waivers and join the Feinberg program after their symptoms did begin to show up, but others couldn't.

Sometimes, the tough men who were the backbone of the recovery effort resisted acknowledging their disabilities. "A lot of those guys didn't want to do the medicals, because they didn't want to get medi-called out," said Harold Schapelhouman, meaning they didn't want to wind up out of work indefinitely. Libretti, for example, continued to work for well over a year after he left Lower Manhattan, struggling through increasingly serious symptoms and ignoring his friends' and relatives' pleas that he confront his condition. When he finally applied for workers' compensation, the construction company he worked for at the World Trade Center site challenged his claim, arguing that his illness could have been the result of jobs he had worked on since.

No one knows how many sick World Trade Center workers failed to apply for benefits at all or dropped out of the compensation process in frustration. For a long while, a worker who inhaled toxic air on September 29 had to go to a different court than one who inhaled it on September 30, due to a complex judicial ruling. "I've got patients who really should have stopped working but they're hanging in there," Dr. Levin told the *New York Times*. "Many of them say they simply are not going to get into a system that puts them through such hassles, even if we advise them to do it for their own protection."[20]

One particularly tragic story was documented by Ridgely Ochs, a staff writer for *Newsday*. Timothy Keller, a city emergency medical worker, drove his ambulance to the World Trade Center in time to be caught in the middle of the chaos when the South Tower collapsed. For months afterward, he returned to the site after his regular work shift to help in the search for bodies. About a year after the attack, his friends noticed that Keller had begun to have trouble breathing. "He would walk two blocks and have to stop. Then that shortened to where he could walk a block and have to stop, and then he could only walk a couple of feet," his 19-year-old son, David, said. But Keller resisted any idea of disability until the spring of 2004, when people began complaining that he was unable to do his job. By then, he had missed the two-year deadline to file for a September 11 injury. His application for workers' compensation, which had to be approved by the city, was rejected under the theory that there was no connection between his job and his illness. His appeal was still working its way through the system when he died in June of 2005, almost penniless.[21]

Early in 2006, James Zadroga, a 34-year-old New York City police detective, became the first city police officer to die as a direct result of exposure to Ground Zero air. His parents said he was diagnosed with black lung disease, mercury in his brain, and pulverized glass in his body. Tests showed he had the lungs of an 80-year-old man. The family tragedy was compounded when Zadroga's young wife, Ronda, died of a heart ailment while he was bedridden and tethered to an oxygen tank. His parents are now raising their granddaughter, Tylerann, who was four when she was orphaned. While everyone presumed that the police and firefighters who came to harm during the rescue efforts were well looked after, Zadroga's parents said the Police Department resisted acknowledging that his labored breathing and persistent cough were job-related and that he was forced to keep reporting for work long after he was too sick to do the job. "I can't pay my bills and work doesn't want to acknowledge that I'm sick, depressed and disgusted," the detective wrote during his last illness. ". . . They remember the dead but don't want to acknowledge the sick who are living." The NYPD, which would not comment on the Zadroga's charges, said that the department now has a policy allowing officers who had worked more than 40 hours at Ground Zero to qualify for disability pension. That law was not on the books when Zadroga, who had worked more than 450 hours at the disaster site, fell ill.

DAVE O'NEAL DOESN'T like television. "I'll just sit there in total silence and do nothing," he says. He has plenty of time to contemplate whether there was a better way to handle the work at Ground Zero. "I think that what they should have done was close down the Ground Zero area for a week, have environmental protection come in and figure out exactly what kind of equipment they were going to need, which I know the government has plenty of money to spend on. And get in touch with Washington and tell them the stuff that we needed to use down there to get the job done properly . . . Whether it took two or three years to do the job . . . it didn't matter. But to have the right equipment, every once in a while send us the right filters and stuff like that."

CHAPTER 9
THE AIR

WHILE THE NATION was still staring, dazed, at the TV replays of airplanes smashing into glass and concrete, Rudy Giuliani started talking about getting New York up and running again. "I think people tomorrow should try to come back to work," he said on September 12. "We really want to get the economy of the city as much back to normal as possible." He'd actually started talking to his inner circle about getting the New York Stock Exchange reopened within hours of its closing on September 11, and proclaimed in a September 12 CNN appearance that it could open the next day.

The desire to return to normalcy, to believe that everything was going to be okay—fast—was nearly universal. While Americans said over and over that September 11 Has Changed Everything, they were simultaneously trying to convince themselves that nothing had changed at all. By September 13, at Giuliani's urging, all the city cultural centers reopened, and by Monday, September 17, the stock exchange and NASDAQ were back in operation. David Letterman, whose late-night show broadcasts from midtown Manhattan, returned to the air on the same day. "I'll tell you the reason I am doing a show and the reason I am back to work is because of Mayor Giuliani," he told the audience. "But in this one small measure, if you're like me and you're watching and you're confused and depressed and irritated and angry and full of grief, and you don't know how to behave and you're not sure what to do and you don't really . . . because we've never been through this before . . . all you had to do at any moment was watch the Mayor. Watch how this guy behaved. Watch how this guy conducted himself. Listen to what this guy said. Rudolph Giuliani is the personification of courage."[1]

The world was indeed watching Rudy Giuliani to see how to behave, and many people would be forever grateful that he had shown them how to be brave. However, the mayor's repertoire was limited. The workers at Ground Zero who had watched Giuliani striding confidently around without a respirator got the wrong role model. And if the people who worked and lived in Lower Manhattan were expecting the mayor to be their champion, they were going to be disappointed.

THE AREA IMMEDIATELY around the World Trade Center included the Wall Street financial district, Battery Park City, and other high-rise apartment complexes that ran along the Hudson at the tip of the island. Some of the buildings were composed of million-dollar condos owned by financiers or corporations who used the apartments as luxurious crash pads for visiting executives. Others were subsidized mixed-income residences that included a large number of old people. Many housed middle-class professionals and their children. Virtually everyone who could leave after the attack did so quickly. Most of the area was without electricity and telephone service—the two Con Ed substations that routed power to Lower Manhattan had been destroyed by the collapse of 7 World Trade Center. And the air was almost unbreathable, laced with an extremely alkaline dust whose corrosive qualities were likened to ammonia.

George W. Bush was no less determined than Giuliani about getting Wall Street to reopen right away. But technologically, that required several days. The stock exchange had its own power supply, but a Verizon office that handled most of the exchange's voice and data circuits did not. Harried telephone officials had to rush in portable power generators. As far as the air was concerned, the federal and city governments had one unified message from the beginning: nobody should hesitate to come to work because of the environment. On

September 13, the EPA was already announcing that it had analyzed its early air samples and "the general public should be very reassured" by the initial results. "EPA is greatly relieved to have learned that there appears to be no significant levels of asbestos dust in the air in New York City," said the agency's head, Christie Whitman. By the 16th, the assistant secretary of labor for OSHA said tests showed it was "safe for New Yorkers to go back to work in New York's financial district," and on the 18th, Whitman announced that she was "glad to reassure the people of New York that their air is safe to breathe."

The Giuliani administration became an EPA echo chamber despite the fact that its own testing had shown hazardous asbestos levels in the air. The city's test results were concealed from the public, according to the New York Environmental Law and Justice Project, which obtained the unreleased numbers from the state and posted them in 2004. No warnings were issued despite the fact that hazardous levels of asbestos were registered up to seven blocks from Ground Zero.[2] The city had taken 87 outdoor air tests between September 12 and September 29. Seventeen showed asbestos levels deemed hazardous by both the city and the EPA. Other tests were classified as "overloads"—meaning that there was so much particulate dust in the collection device that no reading could be taken.[2] These disturbing test results were shared with the EPA. Nevertheless, both the city and the federal government continued to claim that there was no danger.

SOMETHING ELSE THE public did not know was that the EPA's announcements were being vetted by the White House and changed to produce the most optimistic possible tone. On September 13, for instance, the EPA press release quoted Administrator Christie Whitman as saying the agency was "greatly relieved to have learned that there appears to be no significant levels of asbestos dust in the air in New York City." The quote, it turned out, had not been in the original draft. It had been added by the White House Council on Environmental Quality, which went through every public statement in the weeks after the attack, inserting a dollop of cheer here and a sunny adjective there. A September 16 EPA draft reported that dust samples "show higher levels of asbestos," but, after a White House cleansing, "higher" became "variable," and a concluding clause was added: "but EPA continue [sic] to believe that there is no significant health risk to the general public in the coming days." As early as September 13, the subhead on a draft announcing "EPA Initiating Emergency Response" went from "Testing Terrorized Sites for Environmental Hazards" to "Reassures Public About Environmental Hazards."[3]

It was the nation's first test of how high concerns for public health would rank in the age of international terror, and the people in charge flunked on several critical counts. The EPA didn't even have safety standards for some of the dangerous elements, such as dioxin, thrust into the air by the towers' collapse. Its New York regional office was using 20-year-old methods to measure the asbestos fibers, though other regional offices and private testing firms at Ground Zero used far more modern and sensitive detection equipment.[4] The composition of the dust varied greatly from one place to another—even on the same block or inside the same office. But instead of taking that as a warning that what looked safe in one place might not be as harmless elsewhere, the EPA turned things upside down, attributing ominous readings to "spikes" that probably didn't exist anywhere else. Much later, agency officials would acknowledge that Ms. Whitman's famous flat-out announcement that the air was "safe" actually applied only to outdoor air far from the pile, not necessarily offices and homes.

And only to asbestos, not other pollutants.

And only to healthy adults, not children or the elderly.

THE MOST PRUDENT course would have been to hold off any reopening of Wall Street until the picture on the air quality was clearer. Indeed, the day Whitman announced everything was safe, over a quarter of the bulk dust samples that had been collected had shown asbestos at levels above the 1 percent benchmark the EPA would use throughout the post-9/11 period. In addition, the first test results on the amount of lead, dioxin, PCBs, and toxic ash in the air were not even available.[5] At minimum, the government should have been honest with the people it was summoning back to the financial district. The workers were civilians, and they did not deserve to be spun by the people who were supposed to be their first line of defense against dangers in the air. They deserved to know that the results were at best incomplete, and that the area was very probably not a good place for people with asthma or other preexisting respiratory problems to linger. Pregnant women should certainly have been told that the test results were not available on lead, which can lead to developmental damage in children. (Juan Gonzalez of the *Daily News* later reported that 17 percent of the samples taken near Ground Zero showed levels of lead above federal safety standards.)[6]

The amount the experts didn't know about the air was overwhelming. There were elements in the dust that had never been in the air in substantial quantities before, and for which the EPA had no standards for measuring danger. So the agency focused, in the main, on asbestos, something with which it had great experience—although never in a massive outdoor arena and never when the fibers had been pulverized by such an extraordinary impact. Experts examining the World Trade Center dust later would question whether the crashing towers had broken down the asbestos fibers into minute pieces that were difficult for normal instruments to detect and measure. They also argue about whether such tiny fibers were less dangerous than, more dangerous than, or equally as dangerous as those of normal size.

Asbestos is a deadly mineral. Its fibers can lodge in the lungs, and while the effects tend to be delayed, over the long run they can cause mesothelioma and other potentially fatal respiratory diseases. Critics of the way the city dealt with the asbestos and other air problems after 9/11 often compare the response to two events in the recent past. One was a 1989 explosion of a steam pipe in Gramercy Park, an upscale neighborhood near midtown Manhattan. The explosion covered the street and nearby buildings with asbestos-laced mud. As a result, the city evacuated 350 people from those buildings and billed Con Ed $90 million for the cleanup and replacement of ruined property.[7]

The second incident that people pointed to with puzzlement was the discovery that the homes and lawns and gardens of Libby, Montana, had become contaminated with asbestos from vermiculite that was mined in the town. Prompted by a 1999 investigative series by the *Seattle Post-Intelligencer,* the government designated Libby a Superfund site for federal cleanup. After 9/11, Joel Kupferman of the New York Law and Justice Project wrote a blistering essay comparing what happened at Libby to the EPA's performance in New York, arguing that the entire area of Lower Manhattan contained comparable or higher levels of asbestos. One testing from an air vent at a building on Duane Street—a block the EPA said was not even in the contaminated zone—had asbestos levels 50 times higher than the recommended safe levels, Kupfermen found. Though the Law and Justice Project's findings were seen as an aberration by the EPA, no one disputed that the results here and in another study exceeded anything found in Libby.

The difference between those incidents and 9/11 were, of course, size. Libby was a tiny mining community and the Gramercy Park explosion involved only a few hundred people. Lower Manhattan was the residence of about 34,000 and the center of the nation's financial network. The government could shut down Libby or relocate a few hundred Manhattanites, but it did not feel it could do the same for the entire area surrounding Wall Street. In fact, the mere mention of Libby and the dreaded word "Superfund" was enough in itself to send officials into denial. Superfund designation was the very thing the city wanted to avoid. While it would have meant far stricter safety standards for the workers at Ground Zero and far greater protection for the residents and workers in Lower Manhattan, it would have delayed that much-desired return to normalcy indefinitely. The White House and the Giuliani administration were terrified that Lower Manhattan would be regarded as a contaminated no-go zone, sending the financial community fleeing to other parts of the country. "We had been pushing to have it regarded as a Superfund site," said Bruce Lippy, the industrial hygienist, "but people in the political structure said 'we can't talk about it as hazardous.' "[8]

THE WALL STREET workers and Lower Manhattan residents might reasonably have expected their mayor to be their ombudsman, speaking up for them and looking at the federal pronouncements with a skeptical eye. But City Hall did no such thing. Shortly after the attack, Dr. Philip Landrigan of Mount Sinai Medical Center went distinctly off-message when he warned that exposure to the dust and rubble could produce long-term health problems. Giuliani and his health commissioner, Dr. Neal Cohen, quickly disputed Landrigan's assessment.

Landrigan was hardly the only credentialed contrarian. A U.S. Geological Survey team, responding to a White House science office request, flew over Lower Manhattan four times between September 16 and 23, using a sensing device designed for exploring Mars and Jupiter. They also collected 35 dust samples from the ground, finding asbestos, heavy metals, and pH levels similar to those of drain cleaners. Dr. Thomas Cahill and a group of scientists convened regularly by the U.S. Department of Energy took measurements a mile north of Ground Zero and found a level of fine particulates higher than that measured at the Kuwaiti oil field fires during the Gulf War.

Kathryn Freed, the City Council representative for the area, was repeatedly rebuffed by Giuliani's aides when she attempted to get permission to bring an outside testing service to check the air in her district. "Giuliani was in total control and the EPA followed whatever he said," Freed said later. So on September 18, Freed snuck an independent consulting firm past police barricades to secure samples in two buildings inside the zone, finding elevations up to 100 times EPA standards. Six weeks after the attack, when Juan Gonzalez laid out some of the potential health dangers in a front-page story in the *Daily News*, a deputy mayor called Gonzalez's editors to denounce the story and Giuliani went public again with reassurances about how safe it was.

Moderate Republicans from the Northeast tend to be strong on environmental issues. New York governor George Pataki, who had been left looking like a piece of background furniture by Giuliani during the 9/11 recovery, was an activist when it came to fighting pollution and global warming and protecting the upstate wilderness. But environmentalism had never been one of the mayor's strong suits. His commissioner of the Department of Environmental Protection was Joel Miele, who had done a dismal job as buildings commissioner and been rewarded with the top job at DEP. When Miele was hauled before a congressional committee in 2002 to explain his agency's response to the toxic aftermath of 9/11, he threw up his hands, explaining that his agency was primarily "a water and sewer agency" incapable of handling a large-scale hazardous catastrophe. Previous mayors had appointed commissioners with strong environmental backgrounds to the job and even Giuliani had started out in 1994 with Marilyn Gelber, who had earned the respect of environmentalists with her handling of negotiations to protect upstate watershed lands crucial to the city's drinking water supply. Gelber's supporters said she was axed for resisting patronage appointees pushed by the mayor's office. Whatever the reason, just as the Police Department descended from Bill Bratton to Bernie Kerik, DEP sank to Miele, a Queens Republican who kept a revolver strapped to his ankle.

The other member of the Giuliani team on whom the workers and residents of Lower Manhattan were supposed to rely was Dr. Neal Cohen, the head of the Department of Health. Cohen was a psychiatrist who was married to the cousin of one of Giuliani's closest associates, Deputy Mayor Randy Mastro. He had had a quiet and undistinguished reign as the city's top health official, taking center stage only once—in 2000, when Giuliani waged war on the West Nile virus with an intensive spraying campaign. The West Nile virus was a health problem the mayor could relate to, featuring a living enemy embodied in the birds and mosquitoes that were known to carry the disease. With Cohen playing second banana, Giuliani treated the populace to daily reports on the battle. Environmentalists cringed as he ordered spraying in a two-mile radius of each infected bird or virus-bearing mosquito. Spraying crews fanned out over the five boroughs to blanket what the mayor called "hot zones," while critics argued that the health consequences of the pesticides could be far worse than the dangers of the virus. When a woman walking down West 58th Street in Manhattan was nailed by a city truck that sprayed pesticides directly into her face, leaving her with nausea and blurred vision, the mayor reprimanded her for failing to pay attention to the city spraying schedule. (It turned out that no spraying had been scheduled where the woman was doused, but no apology was forthcoming from City Hall.) Giuliani and Cohen assured New Yorkers that the spraying was not a health hazard, while at the same time instructing them to shut their windows, turn off their air conditioners, and take pets inside on spraying night. "There's no medicine that you would take that doesn't carry with it some percentage chance that it could kill you, give you side effects. Those are the kind of choices you are making here," Giuliani explained, ignoring the fact that he was the one making the choices, while the citizens were taking the risks.

The city's workers, including those from the Department of Health and DEP, won plaudits for their around-the-clock efforts after the attacks, but it would have taken the men at the top—Miele and Dr. Cohen—to argue for a more forceful response to the EPA's monitoring of the air quality. Now, as almost always happened in the later days of the Giuliani administration, the men at the top were completely dependent on the mayor for their current stature, and they had neither the standing nor—presumably—the inclination to fight with him.

The mayor deferred to the EPA on the question of air safety because the EPA was saying what Giuliani wanted to hear. But when it came to the question of how the interiors of office buildings and apartments would be cleaned, the city very quickly big-footed federal regulators. The Department of Health took over the job of testing the interiors; Environmental Protection, the exteriors. Having pushed the feds aside, the Department of Health then delayed until November to even start a study of apartment contamination and, though it completed work in December, no results were released until after Giuliani left office. When the study was released nearly a full year after the attack, it revealed that almost 20 percent of the tested apartments "still had interior dust with measurable levels of asbestos." The federal government seemed happy to step aside in favor of the city, even though it would have had strong legal ground if it wanted to take a more aggressive position. The result was a mess. Everyone behaved more or less as though eliminating tons of dust from office towers that were known to contain tons of asbestos, lead, and other toxic materials was just another janitorial function. The cleaning crews contracted for office and exterior cleanups were made up, in the main, of undocumented immigrants, hired by all sorts of different public and private employers under all sorts of different circumstances, with no coordinated oversight by the city. It was just one sign of what an industrial hygienist said was "the Wild West out there."[9]

ABOUT 20,000 RESIDENTS of Lower Manhattan were displaced by the blast. The World Trade Center had been a part of their daily lives. Parents always told their children that if they got lost, they should simply orient themselves by looking for the towers. Their big underground arcades served as the neighborhood shopping mall. "We knew every store in the twin towers," said Dennis Gault, an art teacher and painter. "We shopped in every store down there. They had a place that made pretzels that were delicious and a toy store for the kids." Catherine McVay Hughes, who lived on the 14th floor of an apartment building at the south end of Broadway, remembered how her older son had learned to skate there. "I saw a girlfriend do a folk dance there. It was just kind of, I guess, a town square for the community."

On September 11, Gault and his wife escaped from their 29th-floor apartment, pushing their daughter in a stroller and bearing—improbably—a tutu, which Dennis had numbly thought little Cecilia would need for her first ballet class, scheduled for that evening. Yachiyo Gault, who had worked for the Japanese bank Asahi on the 60th floor of the tower before her baby arrived, was hysterical, convinced dozens of her closest friends were dead. Outside, the esplanade was filled with their anxious, milling neighbors and friends. The Gaults made their way to the ferry to New Jersey and had just scrambled on board when the South Tower collapsed. A giant whitish-gray cloud with pink blotches rushed toward the boat. "I knew the pink stuff was fiberglass insulation," Gault said.

The nearness of the disaster—and the fact that many of the area's residents and workers had seen the bodies of trapped workers as they tumbled to the pavement—meant that returning was a potentially traumatic event. But staying away, cut off from neighbors and schools and everything familiar, was worse. The Gaults, who camped out with friends and relatives until Battery Park City allowed its residents to reclaim their apartments, were unnerved when their daughter began to draw pictures of flaming towers and build models of the towers with her blocks, knocking them down and rebuilding them over and over. But when they took Cecilia to her pediatrician, he told them that they were focusing on the wrong problem. "The drawings are natural. What I'm worried about is the smoke coming out of the wreckage. That is highly toxic."

The plume of smoke did not stop wafting over Lower Manhattan until late December—long after most people had returned to their homes. But even after the smoke disappeared, the dust remained. An estimated 1.2 million tons of it had shot out of the collapsing buildings, and much of it covered not only the floors and furniture of the apartments but also clothing, dishes, toys—the literal fabric of daily lives. Dennis Gault, returning to his apartment after the family's long sojourn away, walked through and saw nothing but the dust. "I thought, here's my two-and-a-half-year-old daughter, and she's breathing toxins, asbestos. We're walking on it. It's in the rugs. I actually panicked."

He turned to his wife and said, "We've got to get her out of here."

They did just that. Gault sent his wife and daughter to live with his in-laws in Japan and embarked upon the most miserable period of his life. In the dead of winter, he moved back into Battery Park City and started to clean. Week after week, he followed the same routine. He got up, went to work, and then returned to his apartment where he would don a mask and scrub and clean. Frightened that turning on the heater would spew more contaminated air into the apartment, he lived and slept in six sweaters and a hat. He threw away all the family's earthly possessions—furniture, clothing, dishes, and cookware.

Most people did not have the resources that got the Gaults through their ordeal. Kim Todd, an actress who lived close to the World Trade Center site, found the dust in her apartment was resistant to all attempts at removal—particularly since her back had been injured when she was caught in the tower collapse. "Every week they'd discover something new about the dust or how to get rid of it," she said. "First it was wet rags, then it was vacuuming, then it was no vacuuming, then it was HEPA filters, but what we discovered was . . . this didn't dust off."

The EPA's own standards would have called for cleaning by professional services that knew about asbestos removal—and which would have cost $5,000 to $10,000 even for apartments of the modest size most New Yorkers live in. But the White House removed that recommendation from the agency's press release, even while the EPA hired a certified asbestos abatement contractor to clean its own regional office, which was located just outside the hazard zone. EPA instead wound up referring the public to the New York City Department of Health for advice. The Department's website recommended that people clean their own homes with wet mops and special high-efficiency particle-arresting (HEPA) vacuum cleaners. (For those who did not happen to have a HEPA vacuum, the Department of Health suggested that "wetting the dust down with water and removing it with rags and mops is recommended.") A press release told residents it was "unnecessary to wear a mask" while cleaning.[10] A city health official told MSNBC Online that those with "underlying respiratory problems" should also follow "simple housekeeping tips like removing shoes, keeping windows closed and changing filters in air conditioners."[11]

The landlords in many complexes cleaned the apartments for their tenants, although very few had cleaning crews with anything approaching professional standards. Others got help from volunteers, including a group of Southern Baptists who had come to the city to join in the rescue and found their calling in going around the neighborhood with mops and vacuums. The insurance company sent a crew to clean the apartment belonging to Catherine McVay Hughes and her family, and then began pressuring them to leave the hotel, where the policy was covering their stay, and return home. Hughes bumped into a FEMA worker, who volunteered to come with her to check out the results. He thwacked the arm of her upholstered couch, which sent dust flying. "He said, 'You have young kids? This is completely not healthy for them. Talk to your pediatrician, and make sure you don't move back until the fires are out.' " Encouraged, Hughes told the insurance company she was waiting and in the end became one of the last families to return to the apartment building. Meanwhile, she dedicated herself to the cleaning project. Ledges that were wiped clean one day, she found, were dirty again the next. She purchased an air purifier with a filter that was supposed to last six months. "It ended up lasting just two weeks and then they were completely filthy, gray with World Trade Center dust."

The EPA's inspector general ultimately concluded that the failure of the government to recommend that residents obtain professional cleaning not only endangered the inhabitants of the apartments but also "may have increased the long-term health risks for those who cleaned WTC dust without using respirators and other professional equipment." City attorney Kenneth Becker offered a creatively convoluted response to the charge, explaining that it wasn't really the city that decided that the residents should clean their own homes: "The owners and the residents always had this responsibility. It was never the city's responsibility to do this and consequently the city could not delegate what it did not have."[12]

FEMA helped pay for the city cleanup of building exteriors when owners claimed they could not afford to do the job, but FEMA officials said that they had been told by City Hall that no formal cleanup program was necessary for building interiors.[13] And the city did not even offer a certification program that would assure apartment building residents and office workers that the places in which they spent their lives had been properly cleaned. The landlords were responsible for cleaning public areas as well as the heating and ventilation systems, and they were, in theory, supposed to test dust samples in their building before it was reopened. "If more than 1% asbestos was found and testing and cleaning was necessary, it had to be performed by certified personnel," Joel Miele wrote in a letter to area residents. However, even presuming the landlord was public-spirited enough to comply with a rule that was very clearly not being enforced, the city never established any procedures to compel them to report contamination and their plans to remove it.[14]

It seems clear that the Lower Manhattan neighborhoods would have done better if the federal government had been in charge of the cleanup, not the city. U.S. District Court Judge Deborah Batts ruled in February 2006 that the Department of Health guidelines "were grossly inadequate," assailing the EPA decision to defer to the city agency as "conscience-shocking." Batts said that EPA's standards were "materially stricter than those the city endorsed" and that the Department of Health guidelines "were meant to apply only to spaces that had been precleaned or tested for asbestos and other toxic substances," though residents were never told that.

There also seems to be ample evidence that if the city had asked the federal government to oversee the indoor cleanup, the EPA would have taken the job. William Muszynsky, the EPA acting regional administrator at the time, said later, "We offered the city assistance on the indoor testing. At the time, we were operating under FEMA and FEMA takes the position that we work in coordination with city officials. The city's emergency response people said, 'We'll take care of indoor testing.'" Asked if he thought in retrospect that the EPA had made a mistake handing this job off to the city, Muszynsky, who left the regional office in 2004 for another EPA assignment, said, "I think the city should have fulfilled what it said it was going to do."[15] Another top EPA official, Steve Touw, added, "If we had to do it all over again, we should have made sure it got done." But legally, he added, it was the Giuliani administration's call.[16]

City Hall never admitted that it told the EPA it did not need help with the interior cleanup, but the EPA inspector general concluded that is indeed what happened, citing "EPA documentation" of conversations with Giuliani officials on September 30 and October 9. The timing corresponded with a series of remarkable statements by the mayor on October 5 that were noted in Fire Commissioner Tom Von Essen's diary. Giuliani is quoted as saying at a cabinet meeting how "pathetic" it is "that everyone is reverting to the tin cup routine," adding that "we need to straighten ourselves out . . . problems will not be resolved by going to Washington . . . every time you go to Washington for help, another big firm says let's get out of New York, they don't know what they're doing." Von Essen observed, "RG runs the city the way I ran my house—you don't take dollars even if someone wants to give it to you unless it is absolutely necessary."

So the decision to dump the job of the most toxic cleanup in history in the laps of already traumatized building or apartment owners and renters was apparently a matter of ideological preference. If ever an American city had experienced a catastrophe worthy of urgent and massive federal assistance—at least before New Orleans in 2005—this was it. From the perspective of the residents of Lower Manhattan, the mayor had picked a very bad time to decide to reclaim his credentials as a fiscal conservative.

The federal agencies would certainly have either run the indoor program themselves or reimbursed the city for the cost of a thorough, professional decontamination program if Giuliani had lobbied for the money. In fact, that's precisely what they did when Mayor Bloomberg finally requested that the EPA take the lead on a cleanup in April 2002. The EPA jump-started the idea, according to a top official, because "over time, we saw that New York City was not prepared to handle all the issues related to indoor air." By the summer of 2003, over 4,000 apartments had been tested and/or cleaned under the program, still far short of what neighborhood leaders demand.

All in all, the efforts in the months immediately following the attack looked like an extraordinarily passive governmental response to a clear and specific health threat. Outside of Ground Zero itself, the main public health responsibility was to give residents enough information to decide when it was safe to return to Lower Manhattan, and then to provide them with competent cleaning and testing services to make their homes free of contaminated dust. Neither the federal government nor the city rose to either challenge.

EPA officials defended their inaction in part by saying that taking responsibility for cleaning dust from all the buildings below Canal Street in Manhattan would have been "a monumental undertaking." The inspector general acknowledged that it would require "a significant effort," but noted that President Bush had told Ms. Whitman that she should "spare no expense and to do everything needed to make sure the people of New York were safe as far as the environment was concerned." In Judge Batts's subsequent decision finding Whitman personally liable, she said that the facts were sufficient to support claims that Whitman had violated the residents' right "to be freeaaaaaaaaaaaaa from governmental policies that increase the risk of bodily harm" when she "consistently reassured members of the public that it was safe for them to return to their homes, schools and workplaces just days following the September 11 attacks."[17]

Of course, Rudy Giuliani's statements were indistinguishable from Whitman's. And almost every failure on the part of the federal government was accompanied by a failure of New York City's leadership to press the case for stronger action. Senator Hillary Clinton held up the appointment of Whitman's successor at the EPA for 45 days in 2003 until she got commitments that the agency would move forward on a more wide-ranging Lower Manhattan initiative. In 2001, Giuliani wouldn't even respond to a health task force formed by the six elected officials who represented the area.

As 2001 CAME to an end, it seemed as if New York City had, indeed, defied the terrorists and come out of its ordeal stronger than ever. Rudy Giuliani was a national hero. The fires were almost out at Ground Zero, where workmen were clearing the site with remarkable speed and efficiency. Everyone expected that a marvelous new tower and a fitting memorial would replace what was lost, along—perhaps—with a whole new neighborhood of shops and housing and cultural institutions. All the glowing reports made the longtime residents of Lower Manhattan feel even more abandoned. Tom Goodkind, a resident of Battery Park City, a huge complex across the street from the World Trade Center, spent two months camping out in a small studio with his wife and children. He listened to the mayor's press conferences, waiting to hear some mention of the residents' plight. "He kept quiet about us and that hurt. We kept wondering why nobody ever mentioned the fact that there were 20,000 people displaced. To this day there's no answer. We were on our own." Even more isolated than the displaced residents were those unable to leave their homes. Councilmember Kathryn Freed was a resident of Independence Plaza, a large mixed-income complex that was without power and cut off from the rest of the world by the security barriers. She felt she couldn't accept offers from friends to get out of the area because so many of her neighbors were trapped. "There were too many people in the building who were homebound elderly," she said. One tenant, a multiple sclerosis victim, could not move on her own, and her home care attendant was unable to get back through the police barricades. Some older residents had to be carried down and taken to shelters. Freed had to struggle to get garbage removal. "I'll always be angry about what the feds, state and city didn't do to help us," she said. "The people responsible for our safety didn't help. It was like—us or Wall Street."

The Goodkind family's apartment was cleaned twice before they returned—once by the landlord and again by the benevolent Southern Baptists. Reassured by the EPA's statements about the air, they resumed their lives until Goodkind developed heart problems and came down with double pneumonia twice. During the second bout, he said, he was lying half-awake, watching the Republican convention, when Christie Whitman took the stage. "I said to my wife, 'Is this an illusion?' . . . I've got double pneumonia and they're cheering her. It was the weirdest feeling."

City Hall's relative indifference to the neighborhood's problems infuriated local elected officials. "The Giuliani administration did nothing but put wrong information on their website and welcomed people to take action that would cause premature death," stormed Congressman Jerry Nadler. Clinton, Nadler, and Republican Christopher Shays of Connecticut, abetted by the Bloomberg reversal of Giuliani's position on requesting aid, finally got FEMA to fund the $80 million Indoor Residential Assistance Program in 2002. When it was finished in 2003—nearly two years after the attack—the EPA concluded that only 5 percent of the residents still had health hazards, down from the 20 percent of 2002. But the inspector general asked "whether the EPA believes possible asbestos contamination in five percent of the residences is acceptable." By the time the EPA finished its testing, the inspector general also observed, many residents had long since returned home. "The full extent of public exposure to indoor contaminants resulting from the World Trade Center collapse is unknown," the IG report concluded.

The federal government planned to resume testing homes and offices, but disagreements stalled that effort in late 2005. Meanwhile, one study showed that women who had breathed the WTC dust and became pregnant had delivered smaller babies, and another reported that asthmatic children who lived within five miles of the site needed more medical treatment after the attack than before.[18] A survey of 2,362 residents found that a year after the attack 43.7 percent reported at least one upper respiratory symptom persisting. Comparing residents of Lower Manhattan with those living far to the north in the borough, researchers found that long after the attack, 21 percent of those who lived near Ground Zero reported their "breathing was never quite right" as compared to 9 percent in the control area.[19]

THE ONE GROUP that completely fell through the cracks during the cleanup was the workers who were hired to clean the offices and, in some cases, large apartment complexes. If residents were given the impression that they could do fine with a damp mop and the right vacuum cleaner, the workers who were hired to work in the dust-covered office buildings were given even less information. The cleaners—often undocumented immigrants—said that they were simply handed mops and told to get to work. "Few of these people we saw were given respirators or taught how to use them," said Dr. Steven Markowitz, who conducted free respiratory screenings from January to March in 2002.[20] And few of them had health coverage.

Dennis Gault, who totally distrusted the competency of the group his building's management sent to clean his apartment, felt sorry for the workers rather than angry. "They mostly looked like illegal aliens," he said. "The sad thing was, I was throwing out all my daughter's toys, and these people were taking them. These were poor people."

In 2005, *Village Voice* reporter Kristen Lombardi interviewed one of the cleanup workers, Alex Sanchez. Sanchez said he spent seven months working in the dust-covered offices in the skyscrapers near the World Trade Center. At 38, he was unable to walk without a cane and was bent over with pain, his breathing compromised by asthma. "The EPA said the air was safe and when you read that coming from a government official you don't second-guess it," he said.[21]

CHAPTER 10

GIULIANI PARTNERS

IN THE DARKEST days of American history, even during the Civil War, elections went on as before. Faith in the orderly transfer of power was faith in the ability of democracy to outlast the worst challenges that human behavior could devise. Yet after the attack on the World Trade Center, Rudy Giuliani began to lobby to change New York City's election rules so that he could remain in office. It was a shocking idea. As stunned and horrified as the city was, it never stopped functioning. Less than one percent of its population was displaced; within a few days almost everyone who worked outside Lower Manhattan was back on the job, and even Wall Street was back operating again soon. The idea that one man was critical to its continued survival was ridiculous. And if America's leaders really did believe that the nation had embarked on a whole new era in which terrorism would be a constant threat, it was the worst possible precedent.

New York was to elect a new mayor in November—the primary, which had been under way when the terrorists struck, was scheduled for a rerun on September 25. Now many people felt reluctant to switch horses at a critical and frightening moment, and Giuliani was constantly being greeted with chants of "Four more years!" as he went around the city. A third term, however, was against the law—a law the mayor himself had supported when it went before voter referendums in 1993 and 1996. And for a while it looked as though Giuliani was comfortable with the idea of letting go. Breaking the rules, he said, would show that terrorism could undermine the democratic process: "What I should do is do the job until December 31, and prepare someone else as the next mayor."

Then his resolve quickly faded. Perhaps he was overwhelmed by the lure of his new role. His public appearances were greeted with thunderous applause and shouts of joy. The president of France called him Rudy the Rock. When he chaperoned the parade of world leaders and luminaries around Ground Zero, there was never any doubt about who the star of the show was. Exactly 10 days after the attack, the *New York Post*—ever Rudy's favorite paper—editorialized that Giuliani should remain in office "for as long as it takes to get the job done." The same day, newly minted "Giuliani for Mayor" posters started appearing around the city. His friends assured the media that there was nothing self-serving whatsoever about the fact that the mayor was taking no steps to tamp down what seemed like a well-organized spontaneous movement to keep him in office. "His expression of being open to continuing to serve is a reflection of his selfless devotion to this city," said his former deputy mayor, Randy Mastro.

Perhaps Giuliani simply felt the candidates to replace him weren't up to the job. There were three serious contenders. Michael Bloomberg, a billionaire businessman who had become a Republican solely in order to get the readily available GOP nomination, had launched a multi-million-dollar advertising campaign, but otherwise had said or done little to give anyone an idea of what kind of mayor he would be. Fernando Ferrer, the Democratic Bronx borough president, was a longtime fixture in city politics. The Democratic front-runner, Mark Green, was an abrasive liberal so despised by Giuliani that he tried to change the charter to strip Green's office of the power to succeed a mayor who died or resigned.

Strategies to keep Giuliani in City Hall abounded. The Conservative Party mayoral candidate offered to run for judge, a maneuver that under the state's peculiar election rules would have allowed Conservative leaders to give Giuliani their party's place on the ballot. Governor George Pataki volunteered that "whatever the mayor's decision is, I will support that decision." Deputy Mayor Joe Lhota called Stan Michels, a Manhattan liberal who'd sponsored a previous unsuccessful bill to repeal term limits. Lhota told Michels that all he had to do was reintroduce his bill, tailored just for Giuliani, and he had reason to believe that the governor would postpone the elections. Michels pondered for two hours and declined.[1] No matter how many plots the mayor's friends came up with, the bottom line was always the same: for Giuliani to gain a third term, state legislative leaders would have to agree to change the term limits law, and their response ranged from cold to icy.

As the rescheduled primary neared, Giuliani said city residents should turn out to vote "if they want to." This lukewarm endorsement of democracy came from the same man who had been exhorting the citizenry to flock to theaters and restaurants to show the terrorists they hadn't won. That night he told Dan Rather he was "open to the idea" of a third term. But neither the Republican-controlled state Senate nor the Democrats who ran the Assembly took the bait. The Albany Republicans generally found Giuliani peculiar and unreliable. For many Democratic members of the Assembly—especially members of the black caucus—he was still their pre-9/11 nemesis. Assembly leader Speaker Sheldon Silver, who was blessed with a political genius for delay, promised to look into the idea.

It soon became clear, even to Giuliani's most ardent admirers, that the third-term scenario was a no go. So the mayor decided to shoot for a consolation prize. Instead of four more years, he would remain in office three extra months—to ease the transition between the old and the new mayor, whoever it might be. The city, he said, is "going to need a lot of help, it's going to need a lot of assistance, and it's going to need politicians who think outside the box. So that all came to me last night that I should start thinking that way also." If he were one of the candidates, Giuliani said, he would have been humble enough to realize that he needed to keep Rudy Giuliani in control for a while longer. "I think that anybody that thinks they're ready for this job on January 1, given the monumental tasks that lie ahead, doesn't understand this job really well," he told the press.

It was a mystery why it would ease the transition to allow Giuliani to stay in office long enough to propose the next year's budget, but not long enough to negotiate its outcome with the City Council. And the biggest question of all was why it would be good for the city—or the nation—to see that a terrorist attack could derail the orderly transfer of power. But the mayor embraced his new idea and went for it with a full-throttle assault that was far more characteristic than the coyness he had been exhibiting over the preceding weeks. His campaign went into high gear when Bloomberg became the Republican nominee and Green and Ferrer qualified for a runoff. The day after the primary, Giuliani summoned Green to an urgent meeting at the ad hoc command center he had set up on the Hudson River piers. There, Giuliani briefly and forcefully explained that he wanted all three of the still-standing candidates to endorse the plan for a three-month extension of his term. Green agreed. Bloomberg, who desperately needed Giuliani's endorsement, also fell into line quickly. But Fernando Ferrer, the last and toughest sell, failed to crack, even when Giuliani raised the threat of finding a way to go for a third term if he didn't get the three months.

For liberal New Yorkers, the drama about extending Giuliani's term was a reminder of everything they had disliked about the mayor before the attack that had transformed him into their hero. "And he was doing so well," sadly wrote *Newsday* columnist Sheryl McCarthy. To Bob Herbert of the *Times*, the candidates' response to Giuliani's strong-arming "may not have shown us who would do the best job running the city, but they sure told us who would be quickest to defend the democratic principles that are the bedrock of the American way of life."

As November approached, it became clear that the state legislature was not, as Giuliani's fans had hoped, going to bow to public outcry and allow the mayor to extend his term. There was not, in fact, any outcry to heed. New Yorkers were following the mayor's advice, returning to their normal lives, and realizing that while their city might be wounded, it was basically intact. Their initial fear of change had faded away. Still, it was not until the last days of the campaign that

Giuliani finally abandoned his own strategies and delivered a critical endorsement to Bloomberg. Reporters noted the mayor's lack of enthusiasm. He arrived more than an hour late for the announcement, spoke in what were for him very muted tones and exited after about 15 minutes. Bloomberg, however, was spending about $350,000 a day on television commercials and, during the last week of the campaign, New Yorkers couldn't avoid hearing that Rudy Giuliani recommended Mike Bloomberg as his successor. When Bloomberg won a narrow victory, many people—certainly including the mayor—felt he owed his unexpected triumph to Rudy.

Bloomberg may have agreed, but he did not invite Giuliani to stay on as de facto mayor for those first three transition months that Rudy had regarded as so critical. In fact, Bloomberg rarely talked to him during that period, perhaps the first sign of the self-confident common sense that defined much of his administration.[2] Rudy's last day in office, December 31, began with five television interviews before 8 A.M., followed by a staff meeting in which his aides took turns thanking him for the privilege of serving in his administration. He attended a fireman's funeral, cut a ribbon for a new police station, rang the closing bell at the stock exchange, and delivered a final positive report on the city's crime statistics. At midnight, he watched the ball drop at Times Square and then administered the oath of office to Bloomberg. He was a private citizen again.

EVEN BEFORE HE left office, the mayor was telling reporters about Giuliani Partners, the management consulting firm he intended to open up with his old City Hall team. The partners were more of a Giuliani posse than a group of peers. Michael Hess, the former city corporation counsel, was named managing partner. His son Geoff would join the firm as well. Richie Sheirer, the hapless commissioner of emergency management, was senior vice president. Tom Von Essen would be a senior partner, as would Bernie Kerik. Tony Carbonetti had worked on Giuliani's mayoral campaign and wound up as chief of staff. Denny Young, the mayor's loyal assistant since their days as prosecutors, had served in City Hall as the mayor's chief counsel.[3] The only partner who came from outside the City Hall crowd was Roy Bailey, the former finance chair of the Republican Party of Texas. Bailey, who'd gotten to know Giuliani when he helped raise money for the abortive Senate campaign against Hillary Clinton, helped finance the new company, which was reported to have a start-up payroll of $10 million a year.

Giuliani had a firm conviction the private sector could learn a lot from his management techniques, but in reality, his record as mayor had been mixed. Spending, which he had pared rigorously in his first years in office, had exploded during his second term, leaving Bloomberg with a budget gap estimated at $4 billion. His attempts to reform the city's school system were generally unsuccessful. The problem of what to do with the city's garbage had become more intractable due to Giuliani's politically driven decision to close down the only local landfill in Staten Island. The city's housing situation had gotten worse. The Fire Department, despite the bravery of so many of its members, was the same inefficient, semianarchic bundle of traditions and union ascendancy that it had been for generations.

But 9/11, as everyone kept saying, had changed everything. The most valuable commodity the new company had to sell was not management expertise, but the aura of America's Mayor. Giuliani Partners initial press releases religiously avoided any mention of the terror attacks—Rudy is described as the man who "returned accountability to city government and improved the quality of life for all New Yorkers." But when their clients, who were very frequently companies in trouble, told the world they had just hired a renowned team of "crisis managers," no one pretended their critical expertise came from handling snowstorms or subway fires. And the partners themselves were far from shy about brandishing the 9/11 club when they were out on the job. Helping the pharmaceutical industry stop Americans from getting their prescriptions filled in Canada, where the same drugs were cheaper, Bernard Kerik would warn that terrorists could be shipping biological weapons across the border in the guise of prescription drugs. Giuliani would make a special phone call to remind a prosecutor about the public-spirited role a client had played after 9/11 or insert plugs for a cell phone client into a talk to public safety officials about what happened to him when the World Trade Center fell.

Giuliani's new offices were located in the Times Square high-rise of Ernst & Young—a building for which he had provided a $20 million incentive package when he was mayor. At the time, critics had pointed out that the site had already been given $236 million in tax breaks over 20 years as part of the Times Square redevelopment project, and that there were plenty of other businesses that could have stepped up to develop the spot without extra incentives. "This is totally unnecessary given the strength of the city's economy," one budget watchdog told the *New York Times*. The accounting giant, which also invested money in Giuliani Partners, was only one of a number of corporations that got financial help from Giuliani when he was a mayor, and did lucrative deals with his firm after he left office. In 2003, for instance, Bear Stearns announced it was putting up $300 million in capital to make investments in the security and public safety sector with Giuliani Partners. (The partners, who were very shallow in financial expertise, were supposed to use their background in criminal justice and crisis management to find good targets for investment.) Under Mayor Giuliani, Bear Stearns had been given $75 million in incentives to build a new office tower when it threatened to leave the city. Critics noted that the investment bank had already received a $40 million tax break when it threatened to leave town during the Dinkins administration, but the deal went through anyway.[4]

Anyone who expected Giuliani to forswear making a profit off of companies that had gotten tax breaks when he was mayor had a wildly overblown vision of the former mayor's code of acceptable business practices. During his only other lengthy stint in private life, after his first unsuccessful run for mayor, Giuliani had taken the world of corporate law as he found it. He abided by the rules, but he did not attempt to transform the culture. It was in sharp contrast to his approach to City Hall, where he entered intent on turning the stale, cynical world of municipal government on its ear and remaking it in his own image. There was something about the world of capitalism that almost always seemed to turn maverick American politicians into docile members of the herd. Giuliani was no exception. He did not take the qualities that made him unique in New York City politics and apply them to corporate America; he took the reputation he had won in New York and rented it out to companies who needed an aura of heroic integrity.

GIULIANI PARTNERS' FIRST high-profile venture was an embarrassment—or should have been if the ex-mayor was planning on becoming more than a hero-for-hire.

Eliot Spitzer, the New York attorney general, was making a name for himself as a politician in much the same way Giuliani had first broken through the public consciousness when he was a federal prosecutor—by taking on Wall Street. Giuliani had been the scourge of the financial community during the high-flying '80s, when he once handcuffed two Wall Street executives and took them on a high-end version of a perp walk. Spitzer was going after Merrill Lynch, which, he claimed, had pressured its analysts to promote stocks they felt were overvalued in order to boost business in the company's investment banking division.

The same morning in the spring of 2002 that Spitzer was going into court to get an injunction against Merrill, Rudy Giuliani telephoned him on the company's behalf. Merrill had not yet publicly announced it had retained Giuliani Partners, and it isn't clear whether Giuliani presented himself as the firm's official representative or merely as the 9/11 mayor who wanted to remind Spitzer that Merrill had bravely kept its offices in Lower Manhattan after the attack. The details of the conversation were never made public, but a source close to Spitzer recalls that "Giuliani did say something about keeping the company downtown, a 9/11 reference." The former mayor raised questions about Spitzer doing this to a New York business, and Spitzer responded that Merrill was defrauding New Yorkers. Then Giuliani floated the argument Merrill had already auditioned—namely, that "everyone in the markets knew what was happening"—and asked if they couldn't "re-start negotiations." After four minutes or so of friendly but stiff exchanges, the former Wall Street buster and the new Wall Street buster hung up.

Spitzer's spokeswoman declined to say much besides "the attorney general noted the interest of investors comes first." What seemed obvious was that Merrill had hoped a quiet word from America's Mayor might encourage Spitzer to negotiate privately rather than via press conference. It was a bad idea on many levels—if Giuliani had been at all introspective, he might have asked himself how he, as a prosecutor, would have responded to an outsider's intervention at such a moment. He might also have contemplated the fact that he was a Republican who was becoming more identified with his party by the minute, asking a big favor of an up-and-coming Democrat who had no reason to feel he owed the ex-mayor anything. In fact, when Spitzer's civil rights division was getting nowhere in trying to obtain police records concerning the city's stop-and-frisk practices a couple of years earlier, the attorney general had called Giuliani twice. He never got a return call.

Spitzer created a national uproar about Merrill Lynch, and by the time Merrill actually announced it had retained Giuliani to help with the settlement negotiations, both Congress and regulators were demanding further investigations. Spitzer's office never heard from Giuliani again and Merrill quickly caved in, paying a $100 million settlement and issuing an apology. Giuliani, who never acknowledged mistakes, denied that anything had gone wrong. "What Merrill Lynch is doing is exactly what they should be doing—using our expertise to straighten things out," he said.

And almost everybody took him at his word. The *New York Times* noted that former treasury secretary Robert Rubin had been deeply embarrassed when it was revealed he had made a similar call to the Treasury Department on behalf of Citigroup during the Enron crisis. "But while Mr. Rubin's sterling reputation might have taken a dent or two, Mr. Giuliani was still wearing his post-9/11 Teflon," wrote reporter Landon Thomas. Indeed, *Consulting* magazine named Giuliani, whose business was still in its infancy, one of the top consultants of the year.

Giuliani's aura was beyond Teflon. Nine months after the attack, a poll by *Family Circle* found that 28 percent of American women picked him as the man they would most like to change places with for 24 hours. Giuliani came in second, immediately behind Tom Cruise. "Who'd have thought he'd have movie star status?" the editor asked Clyde Haberman of the *Times*, who wrote a column marveling at the former mayor's continued wattage. In the months following his return to private life, Giuliani was knighted by the Queen of England and appeared in a Super Bowl commercial in which he was not required to plug any product—just thank America for helping New York. The sponsor, Monster.com, made a $350,000 contribution to Giuliani's Twin Towers Fund. The first anniversary of the attack brought him another huge wave of attention, and it was followed by the release of his book, *Leadership*, which remained on the *New York Times* best-seller list for 25 weeks.

Meanwhile, Giuliani Partners was all over the map, consulting on security for nuclear power plants one day, on efficient bulk purchasing for New York area hospitals another. It signed on to help the troubled, scandal-plagued WorldCom become a "model of corporate governance" and to help Delta Airlines with its bankruptcy. It agreed to review the National Thoroughbred Racing Association's electronic betting systems after a race-fixing scandal. It formed a series of investment alliances that purchased interests in everything from a Tokyo wind-power company to a California firm, CamelBak, which made backpacks with sipping tubes for people like long-distance bikers and soldiers in desert postings. (Bernard Kerik was enthusiastic. ". . . A perfect mechanism to stay hydrated," he told the *Daily News*, envisioning every New York City firefighter equipped with a CamelBak as a matter of course. "If I was a fireman I'd want one.") The Partners also rekindled relationships with some old friends who played central roles in some of the biggest city failures on 9/11. Among them was that "strategic partnership" with CB Richard Ellis, the successor of the firm that had found the city the perfect location for a command center—high above one of the World Trade Center towers. The announcement of the deal, in which Giuliani Partners would be advising Ellis on "location and site assessment" as well as emergency preparedness and fire safety, was made without any discernable sense of irony.

Some of the assignments ended uneventfully. The Greater New York Hospital Association, another entity that frequently lobbied City Hall, hired Giuliani Partners to do a "rigorous independent evaluation" of its bulk purchaser, then renewed the contract anyway. Other enterprises were messy or unsuccessful, such as an initial foray into decontamination, an ironic venture given the get-your-own-mop cleanup of Lower Manhattan. But Rudy appeared to see decontamination as part of his antiterror portfolio, so the Partners formed a joint venture in 2004 with Sabre Technical Services, an environmental company based in Albany. The new enterprise, called Bio ONE, embarked on an expensive and ambitious start-up project, involving a three-story building in Boca Raton, Florida, that had once been the headquarters of the *National Enquirer* and other supermarket tabloids belonging to American Media Inc. During the anthrax scare following 9/11, the *Enquirer* had been the target of a real anthrax attack, in which spores sent through the mail killed a photo editor. The building had been quarantined ever since. Bio ONE intended to demonstrate its skills at decontamination work by cleaning the place up, using chlorine dioxide gas. Giuliani would then fly in to take the first step across the threshold, after which Bio ONE planned to inhabit the old *Enquirer* offices itself, a symbol of the redemption it could offer its clients. The project was expected to cost $5 million. The former home of the *National Enquirer* would soon be "the home of a national hero," Boca Raton's mayor said proudly—though Giuliani clearly did not plan to spend any time in Bio ONE's offices once the symbolic step was made.

Although Bio ONE announced later that year that the building was anthrax free, it never finished the project. The company got caught up in a legal dispute over the ownership of the *Enquirer*'s file of photographs, which Bio ONE seemed to have assumed would be destroyed as part of the cleaning. The threat of litigation shut down the whole enterprise, and Giuliani never got his chance to walk across the threshold. Eventually, the contract between Bio ONE and its owner expired and the building's owner hired a rival firm to finish the job.

Another spate of bad publicity came when Applied DNA Sciences signed up Giuliani in August of 2004, offering $2 million to get advice on marketing strategy for its products—technology that embedded DNA marking in labels to fend off counterfeiters. An article in *USA Today* pointed out that Applied had suffered $35 million in losses from 2002 through 2004 and had no revenue, no cash for operations, and no customers, but that its stock had zoomed after the Giuliani deal was signed.[5] The article quoted Stephen Meagher, a former federal prosecutor, as saying that the situation "has all the markings of something Giuliani himself would have looked into as a U.S. attorney in the old days." (Applied DNA issued a rejoinder saying that the stock rise was due to an announcement that it had secured its first contract, with a Washington state government agency.)

And its "strategic alliance" with Aon, the insurance giant that lost 176 employees on 9/11, certainly raised eyebrows with some of the families. Signed to do "threat assessment" and "pre-incident crisis communications," among other assignments, the Partners actually dispatched Von Essen to an Aon conference in Cleveland to discuss evacuations and emergency planning. Monica Gabrielle, the cofounder of one of the most activist family organizations, the Skyscraper Safety Committee, lost her husband, an Aon executive, on 9/11. "It's an outrage," she said. "What is Aon doing in bed with Giuliani Partners?"

Giuliani wanted to make it very clear that what his company was doing for its money was not lobbying—it specialized in crisis management, not elbow twisting. But many of its most successful ventures involved influencing powers in the Republican-controlled federal government. One of Giuliani Partners' early clients, for instance, was Purdue Pharma, the maker of the pain medication OxyContin, a time-release painkiller that had been proven a great help to people with severe, chronic pain. But OxyContin, which contained an active ingredient more powerful than morphine, was equally popular in the illegal drug trade. It was the drug of choice for right-wing radio commentator Rush Limbaugh, who admitted in 2003 that he was a painkiller addict, and it was widely abused, particularly in rural areas, by users who chewed or snorted or injected it. Small-town druggists were complaining about constant robberies by addicts looking for OxyContin, and the federal Drug Enforcement Agency, a part of the Justice Department, was beginning to blame Purdue's aggressive marketing of the drug for some of the problem. The DEA recommended limiting the right to prescribe OxyContin to doctors who specialized in pain management—an idea that horrified Purdue, for whom OxyContin was a major profit center, in great part because of its popularity with general practitioners.

The drug manufacturer girded for battle. Robin Hogen, a spokesman for Purdue, told a group of public relations executives in 2002 that the company was going to need to "switch over to using more political consultants" and confided to his audience that Purdue was "about to announce next week bringing on a sort of rock star in that area."[6]

The rock star was, of course, Giuliani. While his theoretical mission was to come up with new ways to combat illegal drug use, his real purpose was, as Hogen said, political. Giuliani gave his name and prestige to a brand-new organization dubbed Rx Action Alliance, which swiftly sponsored conferences on pain management and drug abuse, and programs like Dime Out a Dealer which offered $1,500 rewards for antidrug tipsters. He also began making new friends at the Drug Enforcement Administration. Giuliani helped raise money for a DEA museum and appeared with Asa Hutchinson, the head of DEA, at the ribbon cutting for the exhibit. The goodwill initiative was more than successful. Ten months after Giuliani Partners was hired by Purdue to talk the DEA out of restricting OxyContin, it got a $1.1 million contract from the Department of Justice to find ways to improve the Organized Crime Drug Enforcement Task Force, which was charged with, among other things, investigating OxyContin abuse. It was heaven for Purdue—it had employed a consultant to influence an agency that now employed the same consultant to help decide what to do about the issues that concerned Purdue. In the end, the DEA did not limit doctors' ability to prescribe OxyContin, and Rudy Giuliani told reporters he saw no conflict of interest in the situation whatsoever.

Giuliani Partners handled another drug-related project in 2004, when it signed up Pharmaceutical Research and Manufacturers of America. PhRMA was concerned about a very different drug issue—the popularity of drug reimportation. American pharmaceutical companies sold their product at much lower prices in Canada and Europe, where national price controls were in effect. The big profits came in the United States, where Congress had vigilantly guarded the drug manufacturers' right to charge what the market would bear. But American senior citizens had begun taking bus trips to Canada to buy their medication, and, in a far more ominous development for the drug companies, members of Congress were talking about making it legal to import cheaper prescription drugs from across the border. PhRMA wanted Giuliani Partners to prepare a report on the safety of these practices, and although it's possible no promises were made, there was a presumption that the report would find reimportation to be a bad and dangerous thing.

The report found reimportation to be a bad and dangerous thing. "As the nation tightens its borders against possible future terrorist attacks, it risks undermining security and safety by opening them to non-FDA approved prescription drugs," the Giuliani study concluded. Giuliani himself testified before two Senate committees. When the public was invited to take its turn to testify before a federal task force studying drug importation, one of the first speakers was Kerik, who raised the possibility that terrorists could send weapons of biological warfare across the border disguised as prescription drugs.

ANYONE WHO HIRED Giuliani Partners with the expectation of getting much of Rudy himself was likely to be disappointed, given the former mayor's punishing schedule of speeches, campaign stops for favored Republicans, television interviews, and an unending demand that he show up to accept awards from a grateful planet. The partners, or the partners' assistants, provided the actual face time. Giuliani made symbolic appearances—the promised walk across the threshold in Boca Raton or, in the case of Mexico City, the dramatic drive-through.

On the surface, the Mexico City project seemed right up Giuliani's alley. A group of Mexican businessmen, horrified by the city's ever-escalating lawlessness, put up a reported $4.3 million to get Rudy Giuliani to bring his crime-fighting magic south of the border. Who could have been a better choice to lead the turnaround? Announcing the contract late in 2002, Giuliani acknowledged there were differences between New York and Mexico, "but I'm not sure those differences are relevant to crime reduction." Since the differences included the fact that most local police in Mexico City were forbidden by law to make arrests for crimes they didn't witness, that prisons were hopelessly and permanently overcrowded, and that the entire criminal justice system was riddled with corruption, it seemed like a large leap of faith. But in January, Giuliani (who was dubbed not "Rudy the Rock" but the "Iron Mayor" in Spanish) arrived in Mexico City and visited its tough districts in a motorcade that included a dozen bulletproof cars and a helicopter. Critics noted that the caravan ran through red lights and went the wrong way down one-way streets. But the sense that a great presence had arrived from across the border to assess Mexico City's failings and deliver solutions was palpable.

However, the Iron Mayor left the next day. His associates flew back and forth, taking notes and preparing a list of 146 recommendations. Giuliani never returned, though he did do a videoconference for Mexico City real estate managers in March 2003. It was a half-hour monologue that was an amalgam of all the campaign speeches he'd ever delivered in New York, citing Compstat, "broken windows," and every other crime strategy he'd made famous. He even talked about cleaning up the highways between LaGuardia Airport and downtown so visitors would think things were getting better. "Obviously, you're a great politician," said one of the five businessmen who got a chance to ask a question. "But when you talk about the things you managed to do in New York, they seem very easy. They seem as if they were simple and without major complexity."[7] That summer, with something of the air of anticlimax, the Mexican police released the list of 146, saying that they planned to implement them all. The recommendations included the classic Giuliani strategies he'd highlighted in the videoconference. Mexico City's police compliantly declared war on squeegee men, but the effect was different in a city where the squeegees were actually brandished by street urchins rather than indigent adults. The report recommended random, round-the-clock inspections of cabs, which had become flashpoints for criminal enterprise, and the city followed through—although cab drivers complained that the police were focusing on the lower-crime neighborhoods, where they could conduct their searches in safety. In the end, the crime rate didn't drop by anything near the 10 percent anticipated—but then Giuliani Partners had promised that would happen only if all their 146 suggestions were implemented, and many of the suggestions required new laws or constitutional amendments. Two years later, the new police chief, Joel Ortega, announced himself "no fan of Giuliani." Antonio Rendon, a former police official, told the *New York Sun*: "They were not prepared, not at all. They weren't consultants, they were retired policemen."[8]

It was pretty clear that what most of the Partners' clients were buying was political pull and an aura of moral rectitude, no matter what the pleas to the contrary. ("We're not interested in the type of client who might want to come to us and not want substantive work and just want to use the name," said the group's spokesperson, Sunny Mindel.)[9] Broadwater Energy hired Giuliani Partners to help provide "objective security and strategic consulting services" at a time when Broadwater was trying to talk federal, state, and local governments into letting it build a huge floating terminal for liquefied natural gas on Long Island Sound off the New York coast. Broadwater also put a number of former Long Island politicians on the payroll as consultants, despite their lack of security expertise.

Entergy Nuclear Northeast hired Giuliani Partners to evaluate emergency planning and security systems around Indian Point and its four nuclear plants, which were always under the sharp and hostile glare of the communities in which they were located. Then, when Hurricane Katrina struck, Entergy was happy to announce that Giuliani Partners was going "to counsel the company in coping with the aftermath," which included trying to get its New Orleans electric utility back in operation. In making the announcement, J. Wayne Leonard, the chief executive officer, said, "Rudy Giuliani is a proven leader and his team of experts are probably the most acclaimed crisis managers in the world." Not only was New Orleans getting the attention of Rudy himself, the press release noted, but also a team of experts in this sort of catastrophe, such as OEM's Sheirer. All this consulting firepower was not enough, however, to keep Entergy New Orleans from declaring bankruptcy right after the hurricane, or to get power back to large swaths of the city, which were still without electricity months after the hurricane struck.

IF THERE WAS any doubt that Giuliani Partners was all about 9/11, it was erased by the Nextel deal. Nextel was one of the firm's first and most lucrative customers, forming what the Partners called a "strategic alliance to significantly improve public safety communications across the United States." Turning Rudy Giuliani into a leader in public safety communications was a stretch, given the disastrous communications failures at the World Trade Center. Doing it with a company so tied to Motorola, which was once Nextel's biggest shareholder and its principal manufacturer, was even more awkward. The idea of making him the symbol of building decontamination looked reasonable by comparison.

The theme was consistent and clear: "On September 11th, we learned the true importance of interoperable communications. It was a chaotic scene at Ground Zero, but if it weren't for Nextel providing us with interoperable communications tools, it might have been worse." This pitch was so much the core of the Giuliani Partners' message that Richie Sheirer was quoted as repeating it word for word at Nextel-sponsored public safety conferences in Washington, D.C., Austin, and St. Louis in the spring of 2003. The talking point was so tightly scripted that Tom Von Essen made the identical statement at yet another Nextel-sponsored event at the Hilton in New York City. Nextel echoed the point, describing how it had distributed thousands of cell phones at both Ground Zero and the Pentagon (along with mobile cell phone masts to get them working) and how "with local phones jamming and telephone and power main stations down, Nextel's Direct Connect service and two-way messaging remained working throughout the entire crisis, recovery and clean-up." (Bernie Kerik had a slightly different experience with Nextel phones. Testifying before the 9/11 Commission, he said that "as the cell sites dropped, so did the communications. We operated on Nextel, then Nextel dropped. Then they came back up.")

Just as nobody acknowledged anything ironic about Rudy Giuliani's new role as a communications expert, no one admitted there was anything strange about Nextel's attempt to portray itself as the cellular hero of 9/11. But Nextel's role in emergency service communications in New York was far from universally positive. It was, in fact, so problematic you'd have thought that Giuliani would have left office with a bad taste in his mouth for the entire product line. The city had been a big Nextel client when Rudy was mayor, leasing its phones for agencies from the Police Department to the Board of Education to the Office of Emergency Management. But as Nextel phones and equipment became omnipresent, the system developed a stunning, even life-threatening, glitch. The engineering tricks that allowed co-founder Morgan O'Brien to brag that Nextel had managed to stuff 8 million subscribers on a spectrum that was supposed to have a ceiling of 1 million also had a troubling side effect: the infrastructure created interference. In communities across the country, including New York City, Nextel signals were causing those of other users to drop or become garbled to the point of incomprehensibility.

In New York City, Nextel interference affected emergency services well before, and long after, 9/11. NYPD detective lieutenant Ted Dempsey testified in 2000 at an FCC hearing about his department's frustration: "We think we have it right, and along comes a company that puts up equipment and completely wipes out—our problem in New York, we're getting it now, is that my portable radios are pretty much useless below 14th Street. We have a lot of interference issues. I mean Nextel and Motorola are both doing their—making a best effort to take care of that, but how do we go forward building our systems, in whatever technology, whatever modulation, whatever bandwidth, and yet get protected?"[10] Dempsey, who once chaired an FCC public safety subcommittee, said in an interview that Nextel interference in Lower Manhattan was common, beginning in the late '90s and affecting 15 to 20 patrol locations.

While no one can establish that Nextel caused interference on 9/11, there is good reason to wonder. Five days after the attack, a lawyer representing the city e-mailed the FCC about the deployment of all the Nextel equipment at Ground Zero—particularly mobile cell masts—and observed that if Nextel and the city didn't coordinate closely, "it is highly probable that Nextel will disrupt these other critical communications." Another city attorney testified at a subsequent senate hearing that the city and others in a communications coalition held teleconferences twice a day right after 9/11 to "monitor and, if necessary, remedy any interference," focusing especially on Nextel.[11] Even the company acknowledges that its public safety interference increased with the kind of explosions of cellular use that occurred on 9/11.[12] There are also indications that the World Trade Center might have been a particular source of Nextel interference. John Paleski, the president of Subcarrier Communications, which managed sites atop the WTC, said Nextel had several pieces of interference-generating equipment there, including digital and analog antennas. While those antennas were probably knocked out from the moment the plane hit, Nextel apparently had a lot of similar equipment nearby.

In fact, it had so much disruptive equipment in the area that a year after 9/11, it was still a public safety headache. Giuliani led the commemorative services at Ground Zero on September 11, 2002, even while his largest client, Nextel, made the city's interagency communications at the site so "inoperable" during the ceremony that the Bloomberg administration filed a complaint against the company that used precisely that word. Nextel also interfered with communications at 1 Police Plaza that year, and it was cited again and again in 2002 by the city for disrupting public safety in four boroughs, Madison Square Garden, and the city's own Economic Development Corporation headquarters. The Transit Authority filed complaints going back to July 2001. The record was so bad that a Bloomberg filing with the

FCC accused Nextel of causing "debilitating interference." No one in the Giuliani administration had ever said a public word about it.[13]

Giuliani Partners was not hired to defend Nextel's performance, however. It was retained instead to stress how important it was to protect public safety from Nextel interference, and to promote the company's own self-serving solution. Only a few weeks after the attack, Nextel filed a proposal with the FCC, beating its own chest about the need to end its interfering ways. It offered to surrender some of its 800-megahertz frequencies to public safety, cutting the interference. In exchange, it wanted new, continuous spectrum from the FCC estimated to be worth a minimum of $4.86 billion.

Nextel did what it could to orchestrate a chorus of protest about itself—especially from public safety users who operated on neighboring spectrum and were experiencing Nextel-caused dead spots on critical police and fire channels. Who better to stir up those troops than the heroes of 9/11, especially since Nextel was certainly not admitting that it may have caused any interference that day? While Nextel maintained that the Partners were being hired "primarily to market phones to public safety organizations," the *Wall Street Journal* later quoted a senior Giuliani executive as saying that "the main thrust of our work was the spectrum project."

On May 2, 2002, when Giuliani announced his company's partnership with Nextel, the former mayor talked about the interference problem: "Giuliani Partners is committed to helping resolve these concerns and improving the ability of public safety authorities to speak with one another during both day-to-day operations and crisis situations." He also told the *Daily News* that he had reviewed the varied plans for fixing the problem and had decided that Nextel's was the best—"without any doubt."

While support for the Nextel plan in the public safety community was far from universal, many local officials were so desperate to get rid of the interference that they were willing to back anything that seemed to offer a fix. (The Bloomberg administration was concerned about what it took to calling a "plague" of interference and nominally backed the proposal, though it said the rebanding was still "unlikely to eliminate" the plague.) Nextel was also offering to spend a half billion dollars to help cover the cost of all the public safety rebanding that would be needed to make its proposal work, and the $4.86 billion or so in new assets it would acquire in return would be coming from the federal government, not the localities.[14] The company also set up a multi-million-dollar fund, named Richie Sheirer to its five-member board, and reached out to public safety groups, giving, for example, the Association of Public-Safety Communications Officials (APCO) a $3.75 million grant. And when it hired the Partners, it got Bernie Kerik, a member of the Terrorism Committee of the International Association of Chiefs of Police, and Tom Von Essen, a member of the International Association of Fire Chiefs. As part of his Nextel duties, Giuliani appeared at conventions of public safety officials to talk about 9/11 and the need for antiterrorism preparedness. In his keynote speech at the APCO convention in Nashville in 2002, for instance, he delivered his standard talk on 9/11 and his five principles for leadership, one of which was communication. Communication reminded Giuliani of the importance of being able to get through on the telephone in a time of crisis. "I am in favor of your support for the consensus proposal before the FCC that would allow public safety to have more frequencies and better communications," he said. "Thanks to you and Nextel who agreed on that. It can be positive for the future." A trade journal covering the convention called Giuliani "engaging, funny, seemingly honest and informative," but noted that "he never mentioned his consulting company's link to Nextel."[15]

Craig Mallitz, an analyst for Legg Mason, had accurately prophesied: "Public safety is weighing these proposals right now, and I can't think of another guy alive who public safety likes more than Rudy Giuliani. It's probably safe to say, whoever wins with public safety officials wins with the FCC." He was right. After forcing some changes in the Nextel plan, APCO and other public safety organizations rallied behind it, and it was artfully redubbed the Consensus Plan. But Giuliani and the Partners were also making friends at the FCC itself, doing volunteer jobs that had been brokered by Nextel's counsel. The Partners spent more than a year helping the FCC prepare a report about how local government can best communicate with the public during a catastrophe, another odd area of expertise, since Giuliani first tried to get on the air on 9/11 by borrowing NY 1 reporter Andrew Kirtzman's cell phone. Sheirer, Von Essen, and Tony Carbonetti attended a meeting of the FCC's Media Security and Reliability Council just as Nextel hired their firm in May 2002, and another partner, Tom Fitzpatrick, who spearheaded the FDNY's disastrous radio contracting process, moderated an FCC panel.[16] Finally, Giuliani himself was named to a prestigious FCC advisory council two months before the Nextel deal was approved. The FCC ties were another one of those glorious bolts of intersecting interests, much like when DEA decided to hire OxyContin's consultant, Giuliani Partners, to work with its task force.

"Everybody knows what we're doing, so there's nothing hidden," Giuliani told the *New York Times*, in explaining why he had not registered as a lobbyist for Nextel or anyone else. Actually, few did know what he was doing. Giuliani, for example, pushed his bandwidth crusade before the 9/11 Commission and in television interviews without ever acknowledging his interest. Two months before the FCC vote on the spectrum deal in July 2004, Giuliani called for "a dedicated bandwidth for emergency services" on *Larry King Live* and CBS as the solution to the 9/11 communications breakdown. He didn't specifically refer to the Nextel deal, saying only that the bandwidth was "doable but the FCC has to approve it."

By the time Giuliani made these appearances, his consulting deal with Nextel had come to an end, but the partners still may have had the stock options, then valued at $15 million, that they received as part of their compensation.[17] Shortly after the company got its new spectrum without the public auction usually required for such FCC largess, Sprint merged with it in a $36 billion deal. Between 2002 and 2004, Nextel's stock rose nearly tenfold, according to the *Wall Street Journal,* and the *Times* estimated that Giuliani Partners' options were worth $28 million prior to the Sprint merger.[18]

BUSINESS AT GIULIANI Partners was going well, but no one outside knew exactly how well. Giuliani Partners is a private company, and one that keeps its dealings very close to the vest. The company website, which includes the back file of Giuliani Partners press releases, indicated the firm's only public announcement in 2005 was that it had brought on Pasquale D'Amuro, the retiring head of the FBI's New York office, to replace Bernie Kerik. Michael Hess, the managing partner, described the firm's business plan as looking for "big clients with big problems," although the roster would come to include a firm called We The People Forms and Service Centers USA, whose problem was a desire to establish a chain of stores to sell low-cost do-it-yourself will and divorce papers around New York City. Ira Distenfield, the chairman of We The People, said, "Being associated with someone like Mr. Giuliani gives us instant credibility."[19] The outside world could only make guesses at its revenues. Conventional wisdom, encouraged by Giuliani Partners insiders, held that the firm made around $100 million a year, or more than $2 million per employee. (Eric Lipton of the *New York Times* offered a more conservative estimate of "tens of millions" in revenue. Bernie Kerik once reported that in 2002, when the firm was just starting, he had earned $500,000 from the partnership.)

Giuliani had never seemed particularly concerned about money—he wouldn't have been scheming so desperately for a third $195,000-a-year term as mayor if wealth had been his top priority. But his sudden riches came in handy. His settlement with Donna Hanover in the summer of 2002 called for him to pay her $6.8 million over three years as well as child support. Hanover's lawyers estimated that Giuliani's income in 2002 was $20 million, a little more than half from speaking fees and book advances. And he quickly adapted to his new lifestyle, demanding first-class flights and accommodations for himself and his posse when he traveled and purchasing a $4 million summer house in the Hamptons for himself and Judy Nathan, whom he married in 2003. The couple also had an apartment on Manhattan's East Side worth more than $5 million, complete with Rudy's Yankee diamond rings displayed in wood boxes, a lithograph of Winston Churchill above the fireplace, two white Churchill porcelain figures, and a Joe DiMaggio shirt encased in glass.[20] When Richard Grasso, who headed the New York Stock Exchange during the terror crisis, got into trouble for accepting a $140 million compensation package from the exchange board, one of his staunchest defenders was Rudy Giuliani. As mayor, he told Andrew Ross Sorkin of the *Times*, he had not been eligible for a bonus for his work after 9/11, but "if I was and I did a great job, what would be wrong with it?"

Giuliani Partners attempted to present itself as just another vehicle for Rudy Giuliani's fight for justice. "We take these things on if there's good to be done," said Michael Hess. But most of the deals it has made were like the Nextel one—patently about a client's hope to cash in on Giuliani's fame, to borrow a little of his crime-fighting aura, or to make use of the Partners' many connections in the increasingly profitable business of homeland security. Rick Perkal, a senior managing director at Bear Stearns Merchant Banking, told *Newsday* that his company had been impressed by Bernard Kerik's membership on a federal panel that was supposed to give the Department of Homeland Security advice on, among other things, what it ought to be purchasing. Bear Stearns had agreed to invest up to $300 million in new security-related ventures identified by Giuliani Partners, and Perkal said, "Being an adviser in Homeland Security, what has been helpful to us is that he understands the needs of the country. When we look at opportunities, companies that come up for sale, he can say: 'This is a good company. I think it has good growth prospects.' "[21] Although *The New Republic* ran a story in 2004 called "How Rudy Sold Out," all in all, few commentators noted that America's Mayor seemed to have turned into Everybody's Influence Peddler.

THEN, IN DECEMBER of 2004, everything about Rudy Giuliani's postmayoral life—politics, business, war on terror—came together in what initially seemed like one perfect idea: President Bush intended to nominate Bernard Kerik to run the Department of Homeland Security. Kerik, one of Giuliani's closest associates, could take the mayor's ideas and the mayor's influence into the heart of the American war on terror.

Kerik was a classic Giuliani aide, an acolyte who owed his successes entirely to Rudy. A tough high school dropout, he had been abandoned by his mother—a woman who Kerik said he later learned had been a prostitute who was killed by her pimp. He had reason to be proud of the way he had pulled himself up in the world. After a stint in the army he got his general equivalency diploma and joined the New York Police Department, where he became a decorated narcotics detective. But Kerik was still only a third-grade detective, the lowest rank above patrol, when he met Giuliani. He'd never passed a promotional exam. He was 24 credits short of a college degree in a department where you couldn't become a lieutenant without one. His big break—the one that would make him a multimillionaire cabinet nominee—came when he volunteered to work as Giuliani's driver/bodyguard during the 1993 mayoral campaign. The two men hit it off, and Kerik did more than just watch the candidate's back himself. He recruited an entire team of cops to guard him for nothing. After the election, the new mayor appointed Kerik to a job in the Corrections Department, then to corrections commissioner, and then to head of the Police Department in 2000. After 16 months, he left office with Giuliani and joined the new firm.

Kerik repaid all those career-making favors with utter adoration. He remembered that the first day he went to pick up Giuliani, he'd "sweat bullets" because he was so nervous just being around "the most single-minded, brilliant person" he'd ever known. "What he gave me," Kerik wrote, referring to Giuliani's decision to make him commissioner, "was something I had been searching for my whole life. He had unconditional faith in me. He went against the advice of his own cabinet and said something that my mother and father and so many others never said to me: 'You can do it.' "

Like the ex-mayor himself, one of Kerik's main duties seemed to be traveling around the country, giving speeches about police work and terrorism—and, critics noted, putting in a plug for Giuliani Partners clients such as Nextel. He took one leave of absence in 2003 to become senior policy adviser in Iraq, with a mandate to create a new Iraqi police department. But he stayed only four months, and the Iraqi police did not seem to show much improvement. Nevertheless, Kerik was given a speaking role at the Republican convention, and his stock in the Bush administration remained high. Then shortly after the 2004 election he was summoned to the Oval Office, where Bush offered him the Homeland Security job and, according to Kerik, urged him to go in there and "break some china."

Most of the stories about the nomination also mentioned that Kerik had once filed for bankruptcy, that he had been fined by the city Conflict of Interest Board for using police officers as researchers for his best-selling autobiography, and that he had also once sent homicide detectives to interrogate the staff at Fox TV studios when his publisher, Judith Regan, discovered someone had rifled through her purse while she was taping a show. The papers also noted that Taser, a stun-gun manufacturer and Giuliani Partners client, had put Kerik on its board and rewarded him for his 18 months' worth of effort with a shower of stock options—which Kerik had recently exercised to the tune of $5.8 million.

It was a peculiar list of issues, but nothing the Bush administration wasn't prepared to overlook. The president himself was said to be fond of Kerik, a take-charge, go-to-it guy who had been called "Rambo" back in New York, and the mirror opposite of the stolid and much-criticized current secretary of Homeland Security, Tom Ridge. But the nomination was only a week old when Kerik withdrew, claiming he had just discovered that he had not paid taxes for a nanny he had once employed. In fact, Kerik said, the nanny might not even have been legally in the country. Since he had assured White House interrogators that there were no nanny problems in his background, the president's aides were peeved. Giuliani said the situation was "an embarrassment." That was very bad, but it positively paled beside the stuff that reporters had been digging up. The allegations showed up in print hot on the heels of Kerik's retreat:

- As corrections and police commissioner, Kerik had accepted expensive gifts without reporting them as required by city law. The presents included $17,000 to help pay for his wedding reception in 1998, $7,000 of which came from his best man, Lawrence Ray. Ray also told the *Daily News* that he gave Kerik nearly $2,000 to buy a "bejeweled Tiffany badge" and $4,300 to buy furniture for his about-to-be-born daughter's bedroom.

- Lawrence Ray worked for Interstate Industrial, a New Jersey construction company suspected of having ties to organized crime. (The owner told one reporter that he gave Ray, whom he disliked, a six-figure job mainly because Kerik had recommended him.) The company, which also hired Kerik's brother, wanted a license to operate a garbage transfer station in Staten Island.

- Kerik had asked a city official critical to the licensing effort to meet with his good friend Ray, vouched for Ray's character, and allowed Ray to hold his meetings with city investigators in his Corrections Department office.

- Ray was indicted the next year and pled guilty to conspiracy to commit stock fraud.

- Shortly after his marriage in 1999, Kerik purchased two apartments in the Riverdale section of the Bronx, and—although his financial problems had nearly gotten him arrested for failure to pay condominium fees in New Jersey—ordered up extensive and expensive renovations to make one luxurious home. Authorities at the New Jersey Division of Gaming Enforcement, who were attempting to revoke Interstate Industrial's license to work at casinos in Atlantic City, said that the renovations cost $217,800, of which Kerik paid $17,800 and Interstate the balance. Kerik took the Fifth Amendment in the DGE probe. Then the Bronx district attorney opened his own investigation, and soon found himself in the unlikely position of tapping the cell phone of the former police commissioner. Even Rudy Giuliani wound up summoned to appear before the Bronx grand jury.

- After 9/11, Kerik had talked the owner of a luxury apartment building in Lower Manhattan into donating apartments for use by police and fire officials laboring at Ground Zero and then took over a penthouse with a harbor view for himself and used it for trysts with Judith Regan. In a story headlined "Now His Double Affair Laid Bare: Kerik cheated on wife with Judith Regan and corrections officer," the *Daily News* revealed that his simultaneous affair with a corrections officer ended when she found a love letter from Regan in the apartment, called the publisher, and compared notes.

The portrait was so messy that reporters began to question whether there had ever been an illegal nanny at all, or whether Giuliani's friend had just created a relatively mild offense as an excuse to withdraw before the allegations got any worse. "There's a nanny. I swear there's a nanny," Kerik's lawyer told the *Daily News*. The White House gave off-the-record interviews in which they just assumed that Kerik had already been well vetted when he became police commissioner. But the Giuliani administration appeared to have been sloppy at best. After 9/11, when the FBI wanted to do a background check on Kerik so he could receive a very high security clearance needed to see sensitive intelligence information, Kerik never filled out the questionnaire required before the investigation could begin. And the Department of Investigations (DOI), which is in charge of checking out candidates for important city posts, said it could not find the background form that Kerik should have been required to fill out before his appointment as police commissioner.

The same DOI commissioner who'd rebuffed an official referral from the city comptroller to investigate Motorola's FDNY radios, Ed Kuriansky, gave Kerik an apparent pass. Another former federal prosecutor who'd worked with Giuliani, Kuriansky knew a good deal about the Interstate Industrial connection at the time Kerik's NYPD appointment was pending, including its reputed mob ties. The city official Kerik had leaned on to help Interstate was none other than Giuliani's cousin, who ran the city's trade waste commission. Nonetheless, Giuliani maintained he had never heard a word about any of this— something that suggested either that the mayor had not organized his chain of communication very well or that the administration was not populated with officials eager to tell their top boss news he didn't want to hear. Kuriansky, who was known to be especially friendly with Giuliani counsel Denny Young, refused to say what Young or Giuliani knew. "If the former mayor did not know the findings of his own administration's investigative agency, which was headed by his own commissioner, why didn't he?" demanded *Times* columnist Joyce Purnick.

Many of the people who should have raised red flags about Bernard Kerik were now working for Giuliani Partners, advising businesses on matters like how to make sure security systems were well run and employees carefully screened. In fact, Chris Henick, the onetime deputy to Karl Rove, left the White House to join Giuliani Partners in early 2003 and had years of experience with Kerik when the nomination occurred. At first, Giuliani stood by his failed nominee. "It doesn't take away from Bernie's heroism. It doesn't take away from his decency," he said after Kerik withdrew his nomination. But as the bad news mounted, Kerik announced he was leaving Giuliani Partners. He founded his own new company, The Kerik Group, whose website boasted proudly that President Bush had nominated its founder to be head of the Department of Homeland Security.

America's Mayor had selected as America's Police Chief a man whose personal, financial, and social life was a mess, and whose concern for following the rules and avoiding ethical conflicts appeared close to nonexistent. Then he had concluded that Kerik was competent to lead one of the nation's largest and most critically important bureaucracies, an agency with a $7 billion budget. As the Kerik nomination went down in flames—hugely publicized, hugely embarrassing flames—commentators speculated about what the debacle would do to the man who had championed him for the post. But the answer, once again, was that the public seemed prepared to overlook Giuliani's role. A national Quinnipiac poll showed that 68 percent of Republicans—and 45 percent of the public in general—thought Giuliani should run for president.

CHAPTER 11

9/11 ON TOUR

THE INTRODUCTION IS always the same.

"Is there anybody here who is proud to be American?" the crowd is asked. A roar always follows, as it does on this day early in 2006 at the convention center in Birmingham, Alabama. Christian vocalist Staci Wallace belts out "Proud to be an American" and "America the Beautiful." Shots of eagles, mountains, rivers, canyons, and rolling prairies flash by on the big screens hanging from the center of the arena. Finally, there is a shower of red, white, and blue confetti and streamers. A recording of Frank Sinatra singing "New York, New York" is piped through the PA system, as America's Mayor, accompanied by three bodyguards, makes his way to the speaker's platform at the center of the arena.

Rudy Giuliani is the headliner for Peter Lowe's Get Motivated seminar. For all the publicity about his consulting business, his TV appearances, and his increasingly active role on the Republican fund-raising circuit, this has been an important part of his life, and has been for several years. The seminar is a daylong infomercial that moves from city to city around the Lower 48 selling God, country, and ways to make a killing in the stock and real estate markets. He will deliver a variant of the stump speech he gives everywhere. Amid patriotic pomp and circumstance evocative of September 11, he talks about his six principles of leadership: develop strong beliefs, be an optimist, have courage, prepare relentlessly, work as a team, and communicate.

Giuliani may have spoken to upward of a million people in this way, though such numbers are extremely elusive. The balcony seats in the Birmingham arena, for example, are mostly empty. Nevertheless, the local newspaper reports the next day that the arena was "packed" to capacity by 18,000 people. The paper also reports that Jerry Lewis was among those motivating the multitudes, though the aging comedian was nowhere to be seen, having been replaced by America's "premier political comedian," an unknown who entertained the crowd with impressions of past presidents.

Still, a very respectable number of people, perhaps 12,000, are on hand to hear the hero of 9/11. This crowd is largely white and lower middle class, equally committed to evangelical Christianity and the kind of bootstrap capitalism that believes wealth can always be created out of hard work and positive thinking. It is vastly different from the groups who attend events he addresses while wearing his Giuliani Partners hat—"2,000 leaders of the world's corporate real estate industry" in Toronto or "financial industry executives" gathered to talk about identity theft at the Mandarin Hotel in Washington.

One of the great ironies of twenty-first-century America is the power of the lecture circuit. The idea that people could make a living—a fortune—traveling around the country giving talks sounds like something out of Victoriana. But even though people are now up to their ears in audiovisual aids, even though theater is dead and movies are sinking fast, the lure of a single human being giving a talk is still strong. And of all the speakers working today, Rudy Giuliani is among the most popular. His speaking fees have helped to make him an extremely rich man—he commands a normal speaking fee of $100,000, plus expenses. And the events require no particular preparation, since nearly all of his customers are served up the boilerplate leadership speech. When students at the University of Colorado laid out $180,000 to bring Giuliani to campus, some blanched at paying all that money for what were basically canned remarks such as: "Above all, a great leader must learn how to love people, because ultimately you are leading people."

For international events, the price tag generally includes first-class travel and accommodations for Giuliani, his wife, and his bodyguards, and a few charities that engaged him wound up complaining that Giuliani cost far more than their worthy causes collected. Maurice Henderson, the executive director of the Queen Elizabeth Hospital Research Foundation in Adelaide, Australia, said that the foundation cleared only $20,000 from Giuliani's event, while Giuliani had been paid $300,000.[1] A $310-per-plate fund-raiser in Calgary lost money. The charities told a local newspaper that they were disappointed but nonetheless pleased by the publicity they received. "We were kind of thrilled for the opportunity because we really saw a link between the work we do and September 11," said Ruth Ramsden-Wood, president of the Calgary United Way. One of the dinner goers, however, complained that when he bought a ticket to a charity event he expected the money to go to charity. "If you're going to do that, you need to deliver," John Matthews, the manager of a telecommunications firm, told the *Calgary Herald*, adding that he also questioned whether Giuliani should be profiting from the World Trade Center tragedy.[2]

And Giuliani is serious about the money. In February of 2005, he was scheduled to speak at a dinner sponsored by the South Carolina Hospital Association, when the tsunami hit South Asia and the organizers decided to turn their original event, which was a routine fund-raiser, into a special appeal for tsunami victims. Giuliani remained the keynote speaker for the rechristened "From South Carolina to South Asia" benefit, but if the Hospital Association expected him to waive his fee, they were disappointed. Giuliani made a $20,000 donation and flew out of South Carolina with the remaining $80,000, leaving the hospital with a take of $60,000 for the tsunami victims. One Republican state legislator demanded that Giuliani give back the rest. "Frankly his service in New York makes it even more troubling that he would ask for money to appear at an event designed to benefit the victims of one of the greatest human disasters in recorded history," said Representative Tracy Edge of Myrtle Beach.[3] The Hospital Association president countered, "The fact that he is one of our nation's most experienced leaders in dealing with a massive crisis led us to feel like he was a great fit for the tsunami benefit."[4]

The world of corporate get-togethers, conventions, and high-end charity fund-raisers where the head of Giuliani Partners speaks is very different from the Get Motivated tour. The Birmingham seminar, a typical stop, is mostly a vehicle for hawking books, DVDs, motivational tapes, and get-rich schemes. A self-described multimillionaire named Phil Town invites six single moms to the stage and shows them how to get a 51 percent return by trading in eBay stock. Upward of 200 people are impressed enough by Town's spiel to rush to tables where they can sign up for a two-day investment seminar at $1,000 a pop. No one appeared to wonder why a multimillionaire with a proven investment system would be trudging from town to town in the deep South to sell the secrets of his success to strangers.

Get Motivated seminar goers could also sample Bryan Flanagan's four growth stages—unconsciously incompetent, consciously incompetent, consciously competent, and unconsciously competent—and the wisdom of Monster.com founder Jeff Taylor. The grand old man of the motivational circuit, Zig Ziglar, spoke for an hour before raffling off free sets of his motivational programs to three lucky folks. The same programs, of course, could also be purchased at the arena. In *Confessions of a Happy Christian*, Ziglar writes that homosexuality was "the final straw" in the downfall of 90 civilizations. But in Birmingham, the 78-year-old stalwart of the Christian right sticks fairly close to the secular path in his remarks. "Money is not the most important thing in life, but it is reasonably close to oxygen," he cracks.

Giuliani didn't join this team just for the money, and his rather muted delivery in Birmingham suggests he's not in it for the fun. That leaves politics. In his Get Motivated appearances, Giuliani is making direct contact with the base, the Republican voters who will have a large say in determining the party's presidential nominee.

During his postmayoral career, while other potential candidates were in New Hampshire or sending out press releases explaining the finer points of the Bush budget, Rudy Giuliani was touring with Get Motivated, where the audiences were conservative but far friendlier to him and his ambitions than conservative leaders. Giuliani offered in 2005 to speak for free to the Conservative Political Action Conference, a major gathering of conservative activists, but was rebuffed. "I would assume he wanted to come here to boost his conservative credentials, but we didn't think that would be useful," said David Keene, the head of the American Conservative Union.[5]

Rudy Giuliani has long been a supporter of abortion rights, gay rights, and gun control. He could never have gotten elected mayor of New York City otherwise. As a result, many conservative luminaries are cool to the idea of a President Giuliani. "We don't want to nominate the *New York Times*' favorite Republican," was how the GOP's direct-mail guru Richard Viguerie put it, though his antipathy did not get in the way of Viguerie working for the doomed Giuliani-for-Senate campaign in 2000, when the enemy was Hillary Clinton.[6] George Marlin, the conservative New York Catholic who once ran the Port Authority, compiled four pages of quotes from America's Mayor and shipped them to right-wing organizations around the land. A favorite was: "I'd give my daughter the money for it"—meaning, of course, an abortion. It was something Giuliani said to *Newsday* during his first run for mayor in 1989.

It was not just Giuliani's position on the issues that bothered the shepherds of the conservative flock. There were lifestyle questions as well. Rudy was, after all, a thrice-married adulterer who famously moved out on his second wife, Donna Hanover, and into the apartment of two gay friends.

The social conservative movement's foot soldiers, however, seemed far more willing to forgive and forget—or at least ignore. "Gossip," said one woman in the Birmingham crowd dismissively. So beginning early in 2003, Giuliani hit the road on behalf of the Tampa-based Get Motivated seminars. He was often preceded on the bill by Peter Lowe, the tour's organizer. A pop-eyed redhead who has been likened in looks to Howdy Doody, Lowe is also in charge of the spiritual heavy lifting.

"All things are possible to those who believe," Lowe preaches. Audience members are then invited to bring Jesus into their hearts. "Twenty-nine years ago, I made an affirmation, and it went something like this: I said, 'Lord Jesus, I want you to come into my life and be number one at the center of my life,'" he said to a round of applause. Lowe, a 47-year-old Canadian and the son of missionary parents, has been in the motivation business for about 25 years. His first big break came in the late 1980s, when hooked up with Ziglar. His operation got a tremendous boost in 1993 when Lowe persuaded the Great Communicator himself—Ronald Reagan—to speak at one of his seminars. After that, attracting big names seemed to come easily. Gerald Ford, George and Barbara Bush, Henry Kissinger, Charlton Heston, Colin Powell, Goldie Hawn, Joe Montana, Benjamin Netanyahu, Christopher Reeve. The roll call of famous politicians, sports figures, and entertainers seemed endless. Lowe's show moved from city to city, drawing huge crowds at what were then called Success Seminars.

But it all came crashing down in 2001. That year, the motivation impresario decided that Bill Clinton would make a great headliner for the show. How Lowe so badly misread his audience isn't clear. Lowe was making the Great Adulterer the star of his show. The reaction on the right was swift and furious. The show did go on, but as it turned out, the Clinton fiasco was only the tip of the iceberg. Lowe's operation was on the brink of financial collapse, thanks to a soft economy and bad business decisions. The Success Seminar in Chicago in late September 2001 was supposed to feature Clinton, TV host Montel Williams, and former Cubs star Ryne Sandberg. But none appeared, and the venue for the seminar was abruptly switched from the spacious United Center to a roller derby hall. An announcer told thousands of unhappy ticket holders that the changes were prompted by September 11, "so that we would not be in a big place with some of the highest profile speakers in the world." No refunds were offered.

It took chutzpah to use the tragedy as an excuse to mask the fact that Lowe's company was going bankrupt. "The reason the speakers were canceled is that we couldn't pay," said Renee Kirkiewicz, Lowe's vice president for operations.[7] Success Events went under in December 2001. Lowe dropped out of sight for about 15 months following the debacle. He reemerged in April 2003 with a new name, Get Motivated Seminars, the same old show, and a new headliner, Rudy Giuliani.

Lowe's willingness to wave the September 11 flag as a cover for his financial dealings was far more shocking than anything Giuliani or his partners had ever done, but the former mayor did not seem to hold it against him. He embarked on the Get Motivated circuit, speaking to audiences in Boston, Tampa, Detroit, Philadelphia, Seattle, Columbus, Baltimore, Richmond, San Francisco, Washington, D.C., South Bend, and dozens of other cities, large and small.

GIULIANI'S PUBLIC-SPEAKING whirlwind began shortly after he left office in 2002, and proceeded on two tracks. There was a wave of bookings that coincided with the release of his book, *Leadership*. Many of these early speeches were college commencement addresses and highly lucrative appearances before increasingly large audiences curious to have a look at the hero of 9/11. Book sales were boosted by Giuliani's account of how he confided in President Bush about his desire to kill a master terrorist: "I told him, 'If you catch this guy bin Laden, I would like to be the one to execute him.' . . . I am sure he thought I was just speaking rhetorically, but I was serious. . . . Bin Laden had attacked my city, and as its mayor I had the strong feeling that I was the most appropriate person to do it."

The Giuliani on book tour was a sort of Rotarian Rudy, the man who crisscrossed the country talking to grocers, bowlers, college students, physicians, real estate agents, roofers, urban scholars, firefighters, business executives, hospital officials, Knights of Columbus, scientists, farmers, credit union employees, senior citizens, construction companies, computer geeks, speech therapists, corporate security specialists, athletes, and restaurateurs. He was a man who transcended ideology—the same kind of Rudy who, as mayor of New York, had suddenly and inexplicably endorsed Mario Cuomo for governor over the moderate and housebroken Republican nominee, George Pataki.

But when Republican voters go to the polls to nominate someone for president, they really do choose a Republican, and generally a conservative one. If the Get Motivated seminars weren't enough proof that Giuliani understood that, his speech in January 2006 to the evangelical Global Pastors Network in Orlando made things pretty clear. The pastors plan to establish 5 million new churches in the next decade. They believe that to do so will hasten the End of the World and Judgment Day. "I appreciate you. I can't tell you from my heart how much I appreciate what you are doing: saving people, telling them about Jesus Christ and bringing them to God," Giuliani told them. He linked his new religious references to 9/11, saying that his belief in God got him through that day and that "as a Christian, you have more from which to draw." The people who died on 9/11, he said, "came from religious homes where they were taught there is no greater principle than to lay down your life for another."

Political observers regarded Giuliani's appearance as a signal that he was running for president, and tilting his tone. His 2005 endorsement of a Cincinnati Republican candidate for mayor who once wrote that "only born-again believers" should be elected to public office was seen the same way. Having personally told millions of people that the first principle of leadership is a strong set of convictions and beliefs, it was hard to see how Giuliani could retract his multitudinous statements in favor of abortion and gay rights. But he was doing everything he could to make it clear that he was a genuine, tried-and-true member of the party nonetheless. During the 2002 off-year elections he built up a backlog of political IOUs. Elizabeth Dole, Jeb Bush, and GOP hopefuls in at least 15 states got a political pat on the back from Giuliani. He made commercials for 24 Republican candidates and raked in millions for the GOP at fund-raisers here, there, and everywhere. In Kentucky, a reporter described him strolling the grounds of a stately horse farm: "And here is Rudy Giuliani of New York on a torrid summer day in his customary suit and tie, his shoulders in that heavy slump, clapping the shoulders and shaking the hands of 200 people who have donated up to $15,000 a couple to the Kentucky Republican Party to meet him."[8]

Beginning early in 2003, Giuliani made strident support for the war in Iraq an integral part of his 9/11 antiterror credentials. Courage, one of Giuliani's six principles of leadership, became the courage to invade Iraq and topple Saddam Hussein. Just as Giuliani had effortlessly mutated his basic speech to fit the needs of roofers or credit card companies, he now tailored every event to the terror-fighting message. At a hospital fund-raiser in North Carolina, Giuliani spoke of the importance of additional emergency rooms: "We need more of them, and I think the country needs more of them, and they need to be aware of anthrax, smallpox, botulism and sarin gas."[9]

Giuliani's ambitions and Bush's campaign reached a confluence at the Republican presidential nominating convention, which the GOP chose to hold in New York to provide the backdrop to the September 11 theme that would eventually carry Bush to victory. As the national embodiment of 9/11, Giuliani was eager to hand the city's tragedy over to the president's reelection campaign. That wasn't all, however, that the head of Giuliani Partners was doing in a week when lobbyists flooded the city sponsoring events that kept party bigwigs happy and well fed. Pfizer, a charter member of the group fighting to keep cheaper Canadian drugs out of the country, hosted an evening reception at the Rainbow Room in Rockefeller Center in Giuliani's honor.

The first night of the convention bathed the party in the aura of 9/11 and military heroism that specifically wrapped the attack on the World Trade Center and the invasion of Iraq into one seamless package. If polls showed that a large chunk of the American populace mistakenly believed that Saddam Hussein was behind the terror attack on their shores, the convention—and Giuliani—deserved much of the credit. A choir sang military marches to accompany a video showing U.S. soldiers, fighter pilots, and sailors in a fast-paced sequence of martial imagery. The widows of three men killed in the terror attack took the podium with tales of personal grief that moved the delegates to tears. The delegates observed a moment of silence for the victims before New York's City's "singing policeman" capped the tribute by performing "Amazing Grace."

That set the stage for Giuliani. America's Mayor wrapped himself in 9/11 in a 40-minute speech in which he depicted Bush as the warrior president who was avenging the attack, recounting Bush's Ground Zero pledge that the terrorists who brought down the towers would hear from us. "They have heard from us! They heard from us in Afghanistan and we removed the Taliban. They heard from us in Iraq and we ended Saddam Hussein's reign of terror.... So long as George Bush is President, is there any doubt they will continue to hear from us until we defeat global terrorism? ...We owe that much and more to those loved ones and heroes we lost on September 11th." Giuliani had managed, in a few short sentences, to forge the Saddam connection to 9/11 in the American mind that he was uniquely positioned to infer, and that even Bush had, when pressed, conceded was false.

Harkening back to the morning when the Twin Towers collapsed, Giuliani said, "At the time, we believed that we would be attacked many more times that day and in the days that followed. Without really thinking, based on just emotion, spontaneous, I grabbed the arm of then Police Commissioner Bernard Kerik and I said to him, 'Bernie, thank God George Bush is our president.' And I say it again tonight, 'Thank God that George Bush is our president.' "The crowd exploded. Giuliani, who had always specialized in this kind of us-against-them politics, was a smash hit. America's Mayor cemented his position as a frontline fighter in the rhetorical war on terror by addressing 38 post-convention Bush rallies and accompanying the president on a campaign swing to vital states in the closing days. The theme was always 9/11.

Shortly after the president began his second term, Giuliani became a partner in a Texas law firm with blue-chip ties to the Republican Party and red-state politics. Bracewell & Patterson was the firm to go to for energy interests, and it seemed to have developed a subspecialty in representing Republican lawmakers in hot water. Its clients included prebankruptcy Enron, as well as Shell, the Bechtel Corporation, Indian gaming interests, embattled former House majority leader Tom DeLay, and House Financial Services Chairman Michael Oxley, who was facing charges that he and his aides had pressured a mutual-fund group to hire a Republican lobbyist. In 2003, Bracewell & Patterson had been in the news when the firm hired Ed Krenik, a former official at the

Environmental Protection Agency, a day after the EPA—with Krenik's heavy involvement—had made some of Bracewell's utility clients very happy by watering down a clean air regulation.

Patrick Oxford, the managing partner at Bracewell and a major Bush fund-raiser, said he expected Giuliani to show Bracewell "how it is to play in the bigs" and that he understood Giuliani was "keeping his options open" when it came to the presidency. "We are going to support him in whatever he wants to do," Oxford concluded.[10]

Among the other Bracewell partners were Marc Racicot, the former chairman of both the Republican National Committee and the Bush-Cheney reelection campaign; Ed Bethune, a former Republican Congressman (and DeLay's lawyer); and Tony Garza, the Bush administration's ambassador to Mexico. Bethune was one of DeLay's lawyers in the ongoing criminal cases, and the firm had helped DeLay fend off the earlier House Ethics investigation. In fact, Giuliani and DeLay shared the stage the night of DeLay's 2005 indictment, with DeLay introducing Giuliani as a "hero" and first-rate leader at a conservative, pro-Israel banquet in Washington. "Giuliani for President" buttons were visible on the Marriott ballroom floor. The Bracewell firm had also long represented many of DeLay's top corporate donors.

As established as the law firm was in Texas, it was willing to change its name for Giuliani. The new Bracewell & Giuliani opened two offices in midtown Manhattan and began hiring lawyers by the dozens, hoping to make the difficult breakthrough into big-time business and financial litigation in New York. Giuliani was alleged to be dying to get back into court—he was, as a prosecutor, an extremely able courtroom lawyer—but no one explained how he was going to be able to argue cases, run an ambitious law firm, and continue to head Giuliani Partners, plus its offshoots, Giuliani Safety and Security and Giuliani Capital Advisors, while also spending much of his time on the road giving speeches and—oh, yes—running for president.

As he traveled around the country, Giuliani was widely seen as a legitimate presidential contender—the top choice in many presidential preference polls, besting even Senator John McCain. It was often said that Giuliani was flying under the radar, despite the fact that he was turning up everywhere with increasing regularity. He did eschew some of the traditional props associated with presidential candidates. His federal PAC was clearly low budget, and he avoided the customary flesh markets for early-bird White House hopefuls. In March 2006, for example, he was a no-show at the Southern Republican Leadership Conference in Memphis when six presidential wannabes were given 15 minutes to strut their stuff on the runway before an audience of reporters, pundits, and politicos. This strategy enabled Giuliani to largely avoid the scrutiny of national press while he quietly built support within the Republican Party and put together the fund-raising infrastructure that could quickly be used to finance a Giuliani candidacy.

THE MAN WHO strides to the speaker's platform in Birmingham is greeted with a chorus of cheers, shouts, and applause—an appropriate greeting for the man who is, according to at least one national poll, the most popular politician in America. He has grown a bit chubby since his days as mayor. The gaunt, hollow look of prosecutor Giuliani is long gone. The heroes of Giuliani's youth, men like John F. Kennedy and Fiorello LaGuardia, are gone, too. The stars of Giuliani's speech are now men like Ronald Reagan and Vince Lombardi.

Giuliani has done this hundreds of times. He breaks the ice with an anecdote about his first case as a federal prosecutor and then banters with his audience.

"Are leaders born or made? What do you think?" he asks.

"Made."

"That's right. Leaders are made. I think that's a much more important part of it. Of course, it is important that they get born first."

Thereafter, Giuliani's speech is propelled by a parade of cliches and platitudes built around the six principles.

"Think about the captain of a ship. The captain of a ship has to have a destination. If the captain doesn't have a destination, the ship might get somewhere, but no place on purpose. It will be affected by the winds, rains and the . . . a . . . storms. So, to be a leader, you have to have a set of beliefs."

The captain of Giuliani's ship is Ronald Reagan.

"He knew what he believed. And he didn't figure out what he believed based on public opinion polls. He tried to figure out what the right direction for the country was," Rudy opines. Giuliani's manner isn't exactly patronizing, but you can't help feeling that he's speaking to a class that needs to be told things slowly and carefully.

A good leader needs goals, he said. Ronald Reagan had goals. And so did Lou Holtz. Only a couple of weeks before, the politically conservative football coach had confided to Giuliani that the key to his success was goals. At other times and places, Giuliani would cite Martin Luther King Jr. as an example of a leader with unwavering beliefs. But the slain civil rights leader went unmentioned in Birmingham. Also downplayed was Iraq. Rock-ribbed support for the global war on terror and the U.S. invasion of Iraq had become a Giuliani trademark, but by the time he hit Birmingham in February 2006, terrorism and the Iraq war were a brief and muted part of his standard speech.

And so it goes. Tales of 9/11 are mixed into Giuliani's rambling narrative, which lasts about 35 minutes, including the story of a heroic firefighter who ran three miles to reach the World Trade Center. He winds up his speech with a tribute to George Bush, who visited Ground Zero three days after the terrorist attack on the Twin Towers. The remains were still unstable and fires raged underground, Giuliani noted. "But the president came anyway and he stayed there a very, very long time." He makes no mention—in Birmingham or elsewhere—of the main danger at Ground Zero: the toxic air that crippled thousands of police, firefighters, and construction workers who labored heroically in the days, weeks, and months after the attack. The hero of Ground Zero, one might think, was George W. Bush.

"The Secret Service agents who were with him would come up and ask him to leave but he stayed longer and longer and longer and longer," Giuliani reported. Bush did this, for "morale. It gave people a sense that the president cares about us and is willing to take a risk with us."

Giuliani asks the crowd not to give away the ending of the story, which is "just between us." It's the same Bush anecdote Giuliani used in his speech at the Republican National Convention in 2004, but the folks in Birmingham look pleased to be getting some inside information.

In Giuliani's account, man-of-the-people Bush spends most of his time at Ground Zero with New York's rough-and-ready construction workers, who tell him "what to do with the terrorists."

The crowd laughs knowingly.

Giuliani portrays one brawny construction worker regaling the president with numerous unprintable suggestions about what to do to the terrorists.

"I agree!" the president says loudly.

Swept up in the spirit of the moment, the construction worker slaps a bear hug on Bush. An anxious Secret Service agent rushes up to the mayor and says, "Giuliani, if this guy hurts the president, you are finished." The crowd roars with laughter and applause. George Bush really is a regular guy. As an admiring Andrew Sullivan put it after Giuliani's convention speech: "You just cannot imagine a story in which a huge, ham-handed construction worker would ever give John Kerry a big, warm bear-hug. Or that John Kerry would answer a long disquisition from a man in a hard-hat and feel satisfied to respond with two simple words: 'I agree.' Again, Giuliani reminded us of why we tend to like George W. Bush. Personally, I'd rather have pins stuck in my eyes than endure a conversation with John Kerry, but I'd love to hang with Bush."

Or with Rudy.

"I think that man would make an excellent president," Peter Lowe's wife calls out to the crowd as Giuliani makes his way from the arena, to cheers and applause.

PART FOUR

9/11 REVISITED

CHAPTER 12

A SECOND LOOK BACK

When Rudy Giuliani looks back to September 11, he relies not upon the memory of the day itself, but on his memory of the telling of the tale, which he has recounted over and over. That is always the way for people who have lived through a complicated, high-adrenaline event. We sort it out in our minds, assigning order to the confusing rush of images. But there are invariably other realities—sights and sounds and irrefutable facts that we failed to notice at the time, or that we edit out later to give some order to the story in our own minds.

His vision filtered by the years of retelling, Giuliani remembers an order beneath the chaos of falling debris and jumping victims. The city's emergency services were functioning as they were meant to, with him at the helm. "The line of authority is clear," he told the 9/11 Commission. "The mayor is in charge. In the same way the president of the United States is commander-in-chief, the mayor is in charge. That's why people elect the mayor, so they get the choice of whether they get a strong captain or a weak captain or a lieutenant or whatever." Praised for heading toward danger rather than away from it, Giuliani replied, "That was my job. I was mayor. Part of my job description was to coordinate and supervise emergencies. The agencies that were the primary responders were all agencies that worked for the mayor. We had a format for how we did it, and part of that included my being there, so that I could coordinate and make sure everybody was working together."

Rudy Giuliani's performance on 9/11 is legendary, but for most people, the story boils down to one image: the mayor walking north from the disaster, covered with dust. Afterward, in his greatest achievement, he was able to give voice to all the things the rest of us needed and wanted to hear. He articulated our grief, shored up our confidence, and insisted on a levelheaded response that gave no berth to intolerance. We resist knowing anything more—about the eight-year history of error and indifference that preceded that moment, or the toxic disengagement that followed it.

We also actually know very little about what the mayor really did before he stood up, covered in the remnants of the World Trade Center, and began to speak to the world. Giuliani has been allowed to be his own solitary storyteller, and his unexamined 102 minutes transformed him into an international brand of public courage.

There is, however, another version of the day, one of human frailty and confusion. It is a day of real heroes—both rescue workers and civilians, like the Port Authority workers who slowed down their own evacuation to carry along a disabled associate. Rudy Giuliani belonged to a second category—of men and women who behaved as well as they could under terrible pressure, but whose failure to plan ahead caused them to make critical errors at the worst possible moment.

SHORTLY AFTER THE second plane slammed into the Twin Towers, Rudy Giuliani's car pulled up slightly northeast of 7 WTC, where his extremely expensive and ultrasophisticated Emergency Operations Center was perched high up above many large tanks of combustible fuel. Bernie Kerik, who was waiting to meet him, decided it was too dangerous to bring the mayor up to the command center he had so carefully and expensively built. Instead, Kerik pointed out a nearby office building at 75 Barclay Street and said they were "taking people out and setting up a command post" there.

"Is this going to be our main command post?" Giuliani asked Kerik in his own account of the day's events, and Kerik said yes. Then the mayor wanted to know where the Fire Department was set up. Kerik told him that the top chiefs had their command post two blocks away, on West Street, and the two men headed over there.

Looking back with serene hindsight, it's easy to see what the mayor's most important mission should have been at that critical moment. He needed to make sure the proud and fractious police and fire departments were working together. The fire officials were clearly at the center of the action. Chief of Department Pete Ganci, First Deputy Commissioner Bill Feehan, and search and rescue chief Ray Downey had begun the day in the North Tower. (There, they had gotten their first word that Giuliani was on the way, and the tough firefighters laughed a little about the borrowed Secret Service lingo: "The eagle" would soon be joining them.)[1] Then, looking for a location with a better view of the fires, they set up an impromptu command post on the far side of the eight-lane West Street, where they would manage the total incident, working with the board that locates all department resources involved in fighting a fire.

When Giuliani arrived at 9:20, Ganci and the chiefs told the mayor that "they had already gotten some people out above the plane," that they'd been "lucky enough to have a stairway that they could come down." Giuliani thought the chiefs were talking about a stairway in the North Tower, where, in fact, none were ever passable. But he may have misunderstood the chiefs, and they may have been talking about Stairway A in the South Tower, the passageway to survival that only 18 people found. Neither fire dispatch nor 9-1-1, which handled countless calls from people stuck above the South Tower fire, were ever told about an open stairway, though the chiefs apparently knew about one.

"What should I tell people? What should I say?" Giuliani asked.

"The message has to be 'Get in a stairway and come down. Do not stay there,' " the mayor recalled Ganci saying. Of course, the city's emergency operators never stopped giving precisely the opposite advice.

Kerik and Ganci talked briefly. It was the only time the two leaders of these often-dueling departments would speak that day. Uncomfortable about the exposed location, Kerik then said, "Mayor, we've got to get you out of here and set up a command post." Hector Santiago, a member of Kerik's detail, heard the false alarm of a third plane over the radio, and yelled, "Boss, we have to go. There's a third plane coming. We're underneath the building. We have to go." With chunks of the towers falling on West Street, Giuliani urged Ganci to move the command post. They exchanged God-bless-you's.

Then the mayor, Kerik, Deputy Police Commissioner Garry McCarthy, and other top cops all left. The chief of the department, Joe Esposito, was on his way to the fire command post when Giuliani left. Informed by radio that the group was leaving just as he approached, Esposito was also diverted to Barclay Street. Joe Dunne, the first deputy police commissioner, arrived shortly after Giuliani departed and was told to turn around and join the mayor.[2] Deputy Mayor Joe Lhota, who also met Giuliani on Barclay and went to West Street with him, said, "There were no police officials at the command post when we got there and none when we left."

After presiding over endless turf battles between the two proud departments, Giuliani knew how critical police-fire cooperation was and he knew it wouldn't happen automatically. Yet, in *Leadership*, Giuliani wrote: "I turned north and headed to the Police Department command post." In his 9/11 Commission testimony, he said, "I then walked up with, at this point, the police commissioner, the deputy police commissioner, and the chief of the department. I was really brought into 75 Barclay and told this would be our command post."

The "our" was the police and the mayor. Yet the Fire Department was responsible for managing the city response to any fire—a series of interagency directives that Giuliani had signed only a few months earlier said so. Giuliani's role at that moment was to do everything he could to put police and fire commanders at the same post, not participate in setting up a police command post at Barclay that would be separate from Ganci's. If the mayor felt that he needed to go to Barclay—for reasons of safety or to get hard phone lines and hold a press conference—why did he bring all of the top police commanders with him? Why did he never raise the subject of a joint response while at West Street? And since Ganci said he was moving his post, why was there no discussion of a new joint location that would include some of the top police decision makers?

Everyone agrees that a critical problem that day was that the police and fire departments could not communicate; that's one of the reasons the lack of interoperable radios became such a focus of fury. But if the top brass of the two departments were at each other's sides, they could have told each other whatever they learned from their separate radio systems. Many of the command and control issues that might have saved lives could clearly have been better dealt with had Giuliani stopped, taken a deep breath, and pushed Kerik and Ganci to fully and effectively join forces. Insisting that Kerik, McCarthy, Esposito, or Dunne stay at the incident post would have established a joint operation.

"There's supposed to be a unified command where the various individual experts, the chiefs of departments, get together at one command post, not command posts scattered all over the place," said Alan Reiss, the Port Authority's WTC director. "At a unified command post, the various chiefs of the various agencies get together and develop one integrated plan. That's how the system has to work. It's not." Even Tom Von Essen said, "There should be a representative from the Police Department there; there should be a high-level chief from the Fire Department there. They should be controlling the operation from that command post. That day the police did not hook up with the Fire Department. I don't know why."

The National Institute of Standards and Technology found that "functional unified operations were diminished as a result of the two departments' command posts being separated." In fact, said NIST, there's no record that "any senior police department personnel" were assigned "to provide liaison or assist" with Ganci's incident post. Jerry Hauer pointed out the most dire consequence of the split command posts: "Had there been a senior police liaison at the command post, information about what the police were observing in the air could have been relayed to the ground." He, the 9/11 Commission, and NIST agree that at a joint post, the fire chiefs would have gotten the helicopter warnings of collapse.

Rudy Giuliani had the opportunity to make that kind of unified direction happen and, by his own description, the obligation to make it happen, but he didn't. In his first detailed post-9/11 television interview he recalled that he "walked away" from Ganci's post "and took my people with me." But they were not just "his" people, meaning his City Hall deputies. Included in his entourage was the entire police command.

In that same September 22 interview, Giuliani offered a different explanation for his initial decision to go to the FDNY post on West Street: "I wanted to join the Fire Department and the Police Department together at one command post, so I asked where the Fire Department command post was." He had inadvertently described what he should have done, indeed what his own protocol required him to do. But obviously, that story didn't fit the facts. So, by the time he appeared on *The Oprah Winfrey Show* on September 27, he remembered things differently. "And then when I got there," he said, "I wanted to make sure that the police department had a command post so that we could communicate with the White House, and the fire department had one so they could actually focus their attention on fighting the fire and the rescue."

By the time he wrote *Leadership* in 2002, he'd come up with a detailed rationale. He said the separation of command posts was "absolutely necessary" because "the Fire Department had to lead the rescue and evacuation," while the Police Department "had to protect the rest of the city." Since the departments were "performing different tasks," he argued, they had to have different command posts. Of course, the departments have some different duties in virtually all emergencies, but that reasoning not only flew in the face of all modern understanding of how to coordinate these epic catastrophes, but also all the plans Giuliani's own government had put in place. If it were true that different emergency functions required a separation of command, there would have been no rationale for a coordinating Office of Emergency Management. Everybody could just do their own thing. Unified command is now such accepted wisdom that the Department of Homeland Security requires it.

And of course, as the mayor well knew, the Police Department was deeply involved in the rescue and evacuation on 9/11. That's why 23 cops died. Five emergency service units were sent in to climb the steps just like firefighters, as did other plainclothes and patrol cops. Kerik recounts in his book how "our ESU guys were pulling on their masks and marching off toward the buildings" just like the "brave firefighters." And they were doing it from a command post a block north of Ganci's at West and Barclay. Neither Giuliani nor Kerik made any effort even to put those posts together, separated by a couple hundred feet. In addition, the NYPD ran the 9-1-1 operation. It controlled all evacuation in the mezzanines, concourses, and lobbies of the towers. Kerik clearly recognized that he and the Fire Department were conducting a joint operation. He told the commission that once he and Giuliani left West Street, "the key thing for us" was to find "a mechanism to communicate with other agencies."

Giuliani also made the argument that he and the Police Department needed to find hard phone lines and that the Fire Department required clear sight lines as another rationale for the two command posts. He never explained why the Fire Department, disrupted by radio breakdowns all morning, wouldn't have benefited from some hard lines, or why a Police Department with cops all over the towers didn't need to see what was happening. He also never explained why it was impossible to get sight and hard lines in one location—e.g., in any of the buildings across from the towers.

The real, and obvious, explanation for why he left things as they were at West Street was that he was as unnerved as everyone else. The fire and police departments were acting on long-held instinct by staying apart, and the mayor shied away from interfering with men who were busy making life-or-death decisions. It was as human a response as his calming and compassionate statements later that day. But it was also a mistake with consequences, and if New York and the nation actually examined Giuliani's unified command dysfunction that day, both might be better prepared the next time. Unfortunately, admitting all this would not square with *Time*'s salute: "When the day of infamy came, Giuliani seized it as if he had been waiting for it all his life, taking on half a dozen critical roles and performing each masterfully. Improvising on the fly, he became 'America's homeland-security boss,' as well as its 'gutsy decision-maker' and 'crisis manager.' "

There was another reason for the Barclay command post, and Kerik hasn't been as shy as the mayor about mentioning it: security. "I was worried about the mayor and making sure we didn't put him in harm's way," he said later. Kerik's "immediate problem" was finding space "far enough removed that the mayor wasn't in danger." As sensible as protecting Giuliani was, it's a far different explanation than the mayor's rationale for the two posts.

Whatever the mix of reasons, Giuliani has never been forced to explain, by investigators or reporters, how he squares the two-post decision with his own rules for how the police and fire departments were supposed to behave. John Farmer, the 9/11 Commission's top investigator for the city response chapter, says Giuliani can't. "I don't know if he thought of it that day, but yes, it was not consistent with the protocol he established," Farmer says. "I think what he would tell you is that he thought coordination was occurring. He had Kerik with him, and the reality of these situations is that the coordination has to be not just two guys at the top; it has to be more integrated." Asked if Giuliani should be held accountable for this and other disarray that day, Farmer said, "Of course, the answer is yes. If you're the top official, you're accountable."

The 9/11 Commission members reached conclusions similar to Farmer's, but so quietly that no one noticed. The commission report never described Giuliani's step-by-step actions that day, though it chronicles just about everyone else's, and it certainly never mentioned his role in creating two posts. But when it reached its ultimate conclusion that the Fire Department was not "responsible for the management of the City's response as the Mayor's directive would have required," the very next line was "the command posts were in different locations." Thus, the commission's best example of the violation of the mayor's directive was the mayor's own action. "It is also clear," the final report said, "that the response operations lacked the kind of integrated communications and unified command contemplated in the directive," an understated but unmistakable critique of the two-post stumble and more.

The National Institute of Standards and Technology added: "Unified command was hampered by the fire department and the police department setting up separate command posts." It also found that governing Fire Department protocol that day—issued in 1997 when Von Essen was commissioner—said that at a fire like this, "the departments act as 'one organization' and are managed as such." Instead of "several posts operating independently," the department circular provides, "the operation is directed from only one command post." Daniel Nigro, the only top chief at West Street to survive, said, "I think there should have been one command post. It should be run according to the incident command system, and that system puts one person in command and all the other agencies are there and they work from a single location." Anthony Fusco said the command structure that morning was "really a breakdown in coordination," attributing Giuliani's departure from West Street with the police to "the kind of control the NYPD has" on mayors.

Pete Gorman, the president of the Uniformed Fire Officers Association, said at a City Council hearing in 2002, "The NYPD and the FDNY performed heroically on 9/11, but obviously operated parallel and separate from each other. How many firefighters and police officers must die before someone addresses this problem?" Ray Kelly, the police commissioner who preceded and followed the Giuliani years, said in an interview, "Sure, the separate command post was a violation of the protocols. The radios would have been no problem if they had been at the same command post, if they'd been face-to-face. The Office of Emergency Management was supposed to make that happen under the protocols, but Jerry Hauer wasn't there any more. OEM had the power to direct that to happen. Giuliani had the power to direct that to happen."

THE MAYOR'S MAIN mission, as he has put it in repeated accounts, was to gather the information he needed to tell a television and radio audience what they should do, especially people in jeopardy. By the time he talked directly to an audience, however, both towers had collapsed, and the message Ganci asked him to give occupants was moot. The mayor was, in the end, just one more dispatcher who failed to relay useful information. He said he went to Barclay for hard phone lines, but once he got there, his attempt to speak to the vice president went nowhere—someone's phone line went dead, although it's unclear whether it was Giuliani's or Dick Cheney's.

The one call the mayor did complete there—to Karl Rove's deputy, Chris Henick—was pointless. He asked Henick "if we had air support" and Henick said it had been dispatched 12 minutes earlier, which would mean about 9:40. It was more a nervous query than a serious request for information. Both Kerik and Tony Carbonetti had already told Giuliani that they'd reached Washington and been given air support assurances. In fact, he'd already seen the evidence. When Kerik, Giuliani, and the rest of their group were walking back from West Street to Barclay, Kerik remembered, "We hear a jet coming and someone yelled, 'Incoming.' But we realize it's a friendly."[3] (Henick's information was actually wrong. Fighter jets from Otis Air Force Base in Massachusetts had been dispatched at 8:46, maintained in a holding pattern off the Long Island coast for a considerable period, and were finally ordered to fly over Manhattan at 9:13. They were the only air support dispatched for New York.)[4]

Right after the Cheney call disconnected, the South Tower collapsed, and Esposito yelled for everyone to "hit the decks." No one in the Police Department had apparently considered how Giuliani and, by then, a very large entourage would get out of the building in an emergency. So, when the tower knocked out windows and drove rubble and ash into their first-floor safe haven, the group ran through the basement until they found a way into a neighboring building and out onto the street.

As they then walked up Broadway, and then Church, finally hooking up with cameras and press, another plane passed at around 10:20. "We heard a plane overhead, and someone said, 'We're being attacked again,' " Giuliani told Tim Russert. "And then someone else said, 'No, it's one of ours.' And I remember registering at the time, this is unreal.

'That's one of ours.' And it turned out it was, and it was a great sense of relief that it was one of ours." Sheirer said it was "like something out of the movies." If so, it was a rerun, since Kerik and Giuliani had already seen air cover—at least 45 minutes earlier. The fact that Giuliani conflated or confused the two events was meaningless—except as an example of the fragility of memory in a moment of crisis.

Von Essen finally joined the mayor's entourage on this now-famous walk north. Giuliani had begun trying to get Von Essen to join them well before the first collapse, telling Carbonetti he wanted his police and fire commissioner with him. But Von Essen roamed from West Street to 7 WTC, and never found the roving band of brass until after the South Tower disintegrated. His lost half hour was a disconcerting example of just how bad the line of communications between the two departments was. Kerik's entire top echelon was with him at Barclay, as were his and the mayor's details. Two Fire Department aides were with Von Essen. Everyone was outside, where radio communications weren't as difficult as they were through floors of steel and concrete. Yet no one at the top of either department could talk to anyone at the other. No one even carried the 800-megahertz interoperable radios the Office of Emergency Management had given both departments. So the fire commissioner couldn't find the Police Department's overall command post when he was within a couple of blocks of it, even after he was summoned by the mayor. In the end, Von Essen simply stumbled into Giuliani on Church Street, when everyone moved north after the first collapse.

Had Von Essen made it to Barclay when Giuliani did, some semblance of a joint command post might have taken shape for the 10 to 15 minutes they were there, though fire commissioners do not function as operational chiefs. Since Kerik's aides were monitoring the aviation channel, Von Essen might have heard about the warning of the South Tower's imminent partial collapse. Von Essen's aides could then have radioed that information to the chiefs. All the major studies of the events of 9/11 have stressed how lifesaving that missed warning might have been—especially to responders and civilians in and near the South Tower.

After finally picking up Von Essen, the mayor, of course, wound up covered in soot, searching again for a command center, pointing people north. The group with him broke into a firehouse to establish a post, then decided it was too close to the targets. Next, they picked the Tribeca Grand Hotel, sent cops ahead to secure it, and stood inside for a moment, leaving only when they realized it was all glass at the top. Even Giuliani eventually acknowledged that it might have been a mistake that his entire 25-member inner circle, including three deputy mayors, the police, fire, and Office of Emergency Management commissioners, was marching with him on this hazardous pilgrimage, a vulnerability that hardly reflected strategic thinking. This time, Giuliani's preference for the comfort of a huge entourage had disconnected the city's management and its fighting force at a crucial moment.

The only time this confounding management choice took the form of a critical media question was on Fox the day after Giuliani's commission testimony in 2004, when John Gibson asked Richie Sheirer about it. Gibson referred to "the worry" about how the Giuliani entourage had operated, questioning whether it was "fortuitous" that a single "chunk of concrete" hadn't fallen "on Rudy Giuliani, you or somebody else," causing "the whole thing" to have "fallen apart." Gibson appeared to be questioning the wisdom of the fact that "all of the leaders of the city's emergency structure got together and had this little command center that moved around." Sheirer's answer was pure bluster. "No, there was nothing fortuitous about it," he said. "It was well planned. Our succession plan for the highest levels of government, the mayor and people like me, is very well in place and embedded. That was implemented to the degree that it needed to be."

Kerik was actually a prime example of this managerial dysfunction all morning. For the 102 minutes when the city most needed a police commissioner orchestrating an overall response with an embattled Fire Department, Kerik became Giuliani's bodyguard, just as he had been in the 1993 mayoral campaign. His own account of what he did that morning contained no indication that he was actually managing the police response to this emergency. The command center at 1 Police Plaza wasn't opened until 9:45, an hour after the attack, a decision that led the McKinsey study to raise questions about why it was "underused."

McKinsey also criticized the "number and continual movement of command posts," and the absence of any "clearly identifiable, main command post," all errors associated with the top brass including Kerik, who, unlike Von Essen, is an operational chief. "Many leaders of the Department," the independent consultant found, "indicated that they operated primarily from instinct and experience during an emergency rather than according to a prioritized or structured set of objectives." Only 45 percent of the 557 cops who were surveyed by McKinsey said they "received clear instructions regarding my role on 9/11," with 34 percent saying they didn't and the rest undecided. A meager 24 percent said they were "confident" that the Police Department had adequate emergency plans. Remarkably, 89 percent had no training in building collapse, 84 percent had none in counterterrorism, 73 percent none in fire rescue/evacuation, and 70 percent none in bio/chem. Of the few who had training in any of these areas, less than a third found it "useful."

The report, which commended the NYPD's Emergency Services Unit and other police divisions, faulted virtually everything Kerik did that day without naming him or anyone else in top management, citing:

"• Perceived lack of a strong operational leader commanding response

- Unclear roles and responsibilities among some senior leadership

- Large proportion of NYPD leadership responded to incident site and were therefore at risk

- Absence of clear command structure and direction on 9/11 and days after, leading to inadequate control of NYPD response

- Many field commanders operated independently of one another and of higher levels of command

- No central point of information regarding incident, with leaders acting largely on personal observations."

Instead of dealing with any of these complex tactical issues on 9/11, Kerik's decisions—at 7 WTC, West Street, Barclay, and the basement—all revolved around the mayor's safety. Chris Marley, the building engineer at Barclay who guided the Giuliani group out, said "Kerik had his arm around the mayor to protect him." Kerik was later asked what his priorities were that day and he responded, "Protecting the safety of everyone in New York, your safety and protecting the mayor." He told NPR, "Well, the first thing to do was to get the mayor out of there and get to a secure site." With Kerik, Esposito, Dunne, and McCarthy guarding him at points, Giuliani was protected by the highest-ranking detail in the history of the New York City Police Department. Yet not once did he look around and ask the question: who's running the shop?

When asked who was in charge of the NYPD that day, Kerik's successor, Ray Kelly, who has studied the police response as closely as anyone, said at first that the chief of the department was. But the chief, Joe Esposito, was with Giuliani at crucial moments as well. While Esposito later said he was "making sure" there was a "coordinated" response, he never spelled out what he did to achieve that, and everyone agrees it wasn't done. Esposito's other two purposes were identical to Kerik's. He said he "wanted to make sure the mayor was safe" and to inform Giuliani "of exactly what we were doing to address the attack." Esposito's most memorable reported act of the morning was telling the mayor to hit the deck at Barclay. The other top police professional, Joe Dunne, admitted later that he "probably shouldn't have been there," adding that "we do have to become more disciplined."

"I don't know de facto who was in charge," Kelly said. "The police commissioner was the head of the organization. I don't know who was directing. I literally don't."

Kerik was with the mayor because Giuliani wanted him to be. "I need the police and fire commissioners with me," Giuliani said when he summoned Von Essen. He also reached out to Richie Sheirer—the third member of the team who would be at his side for every 9/11 press briefing, then go with him to Giuliani Partners. All three had no real management credentials until Giuliani promoted them. Von Essen and Kerik went from the lowest ranks of their departments to the very top without ever passing a promotional exam. Giuliani had begun his mayoralty with a circle of managers, like Bratton and Hauer, with track records elsewhere. He was ending it with a cult of personality. When he chose Kerik over the seasoned professional Dunne, he told reporters that the decision had come to him in a moment of personal inspiration. Not surprisingly, all campaign bodyguard Kerik could think about in a moment of great crisis was protecting the leader, even if it meant leaving a void in the department he was charged with commanding. To him and Giuliani, it was a matter of duty, respect, and gratitude.

Despite all these missteps, Giuliani was depicted almost immediately as the calmest man in the eye of the worst storm—decisive, self-sufficient, ironhearted. "It was so well orchestrated that you would have thought he had prepared for it forever," his lifelong secretary Beth Petrone-Hatton told *Time* magazine. His own *Time* comments set the subsequent television interview tone: "There were times I was afraid. Everybody was. But the concentration was on. If I don't do what I have to do, everything falls apart. Something I learned a long time ago, from my father, is that the more emotional things get, the calmer you have to become to figure your way out. Those things have become a matter of instinct for me at 57 years old. I didn't have to invent them." He told CNN, "When it's an emergency, I'm very, very calm and very deliberate."

If Giuliani had actually been doing all the things he now sees himself as having done that day—prioritizing, making strategic decisions about deployment of personnel, command centers, and communications—it would have been a superhuman performance. But actually, in those first hours, Giuliani was doing what most of us, in his place, would have done—struggling, stumbling, and even making a weighty mistake, in the case of the two command posts. His decision to try to get on the air as quickly as possible was sensible, as was his hunt for phones and, later, an alternative command center. But as unforgettable a visual as he was, roaming the canyons of Lower Manhattan, he did not do one thing in those 102 minutes that had any impact. Maybe no one could have. The most horrible thing about 9/11, in retrospect, was the inevitability of so much of what happened, with many fates sealed by the floor you were on when the crash came.

Since Giuliani's legend is tied so inextricably to his mere presence in the dangerous and dusty streets of Lower Manhattan, he has on occasion led people to believe that he intended to go to the perilous scene of the disaster all along, rather than the command center. A few pages after writing in his book that he was headed for the command center, he suggested the opposite. "After September 11, people would tell me that it was brave to go to the scene of the attacks," he wrote. "It was actually just carrying out my usual practice for any significant emergency." Kerik, Carbonetti, Lhota, Von Essen, Dunne, McCarthy, Esposito, Mike Hess, and half a dozen other key staffers established by their movements that the command center was the instant front line of defense. That's where they went to find Giuliani, who arranged to meet Kerik in front of it. Lhota, the first to tell Giuliani about the attack and his highest-ranking deputy mayor, says that he waited outside the command center with Kerik "and fully expected to go in it with the mayor when he got there." In suggesting that he went to the scene, not the command center, Giuliani is attempting to meld the two places in the public mind, turning the disastrous siting of the center into a rhetorical benefit. If the command center had been in a sensible location in Brooklyn, where Jerry Hauer wanted it and Mayor Bloomberg put it, Giuliani would not have even been in Manhattan.

That's not the only way he's added to the legend. His memory now includes facing down great physical danger. "I didn't have time to be afraid, Oprah," he said on the Winfrey show shortly after 9/11. "As I look back on it, I realize how dangerous it was." Asked by Barbara Walters if he "thought he was going to die," he said: "Now I do . . . when I think back on it." On another day, the frightening circumstances inside 75 Barclay would have been a dramatic event, with the mayor and his posse rushing from locked door to locked door in the basement, searching for a way out. On 9/11, however, with everything tens of thousands of people went through, it was a blip on the screen. Thousands of others faced smoke and dust and debris on the street, not in the relative safety of a sturdy and off-site building, far removed from the Twin Towers. The Department of Buildings assessment of 75 Barclay was that there was "no interior or exterior damage to the structure observed." The janitors who helped Giuliani through the basement stayed in the building long after he left and were even at the corner later when 7 WTC fell. The engineer, Marley, who gave his face mask to Giuliani, says that 7 WTC, which collapsed seven hours after Giuliani left the Barclay building, "did the most damage" and "felt like it fell right on us," but even then it was just "mostly broken glass."

And it isn't just his own story that he's hyped. Giuliani has repeatedly contended that 25,000 people were rescued, though government investigators determined that there were actually 15,000 survivors and that most of these people were able to make their own way to safety.[5] While these facts do nothing to dim the magnificent bravery of the firefighters, police officers, and other responders who saved many lives that day, they do turn Giuliani's claim into just one more self-serving boast.

THE CENTERPIECE OF Giuliani's experience on 9/11, his dust-covered march uptown, was truly important to the city and the nation. His ordeal was not about management or even leadership—it was the sight of the mayor sharing that terrible experience with so many other fleeing New Yorkers. The symbol of the city was on the ground with his constituents, dirty and determined, conscious of the fact that there were many others who had been less fortunate. He did not have to save any lives to be important that day. Imagine how different our memories of Hurricane Katrina would be if Mayor Ray Nagin had been out in the water with the dispossessed, splashing his way toward the Convention Center.

We rely on our leaders to behave well in such a moment, to set an example of calm and compassion. But we do not expect them to manage the intricacies of the rescue operation. For that, we hope there are men and women throughout the government who have been preparing and training just so that if a crisis comes, they can operate on instinct, yet automatically make the proper decisions. If the mayor of New York had made sure that the city's emergency headquarters was securely located and had put in place communications and command systems that worked, he would have been of greater service on 9/11—even if he had spent the whole day cowering under his desk.

Giuliani has never acknowledged a single failing in his own performance. Yet he did nothing before September 11 to alleviate the effects of a terror attack. He embodied his city's lack of preparation on West Street that morning. And he did not do anything later that matched the moments of grace and resolve he gave us the day we needed him most. What we have left is this: at a moment when the public needed a hero, Rudy Giuliani stepped forward. When he assured New York that things would come out all right, he was blessedly believable. It was a fine thing. But it was not nearly as much as we, at the time, imagined.

ACKNOWLEDGMENTS

WE SIGNED THIS book in 2002, and it represents nearly four years of work. There are three extraordinary people who must be thanked at the outset. Anna Lenzer has worked as our research assistant for a year and a half, supplying us, sometimes seven days a week, with new reporting that was always reliable, insightful, and resourceful. There are few passages in this book that she has not contributed to, and her understanding of the technical aspects of the communications issues raised in the book has greatly informed those chapters.

The other two indispensable people are our wives. Fran Barrett and Gail Collins have not just supported us through tough times and long, unpaid, leaves from our jobs at the *Village Voice* and CBSNews .com, they have bridged the differences between us with skill and care. Without their mediation and perspective, we would have spent more time struggling with each other than with the story of this remarkable slice of American history. Blessed are the peacemakers.

David Hirshey of HarperCollins signed the book all those years ago and guided it through its several evolutions, always true to the story and a believer in us. Nick Trautwein joined him in the editing process, helping us figure out how to hone the message and rescue the drama from the detail. A book, of course, is just a writer with an idea unless someone can convince a publisher to buy it, and no one is more convincing than our agent, Alice Martell. She did a magnificent job for us.

There was also a special group of supporters who combined to pay Anna Lenzer's salary and other research costs: the Fund for Investigative Journalism and John Hyde, The Nation Institute and Hamilton Fish, and the Community Service Society and David R. Jones. Their grants underwriting the research were actually made to City Limits and The Center for An Urban Future, the nonprofit organizations that acted as fiscal conduits, and special thanks are due Jennifer Gootman, Jonathan Bowles, and John Broderick, who made that possible and handled the financial records.

At CBSNews.com, Mike Sims and Betsy Morgan were extremely gracious and understanding bosses who readily granted Dan Collins a leave of absence to write the book. Christine Lagorio did double duty as a researcher for the book at both CBSNews.com and the *Village Voice*. Elizabeth Harris and Karen Avrich also provided valuable research assistance.

The *Voice* was also a special supporter of this project. It allowed Wayne Barrett to take a collective 13 months off, spread out over three years, to devote himself to this book. Editors Don Forst and Doug Simmons, as well as publisher Judy Mizner and new owner Michael Lacey, supported these leaves. They also allowed Anna Lenzer and an army of interns to work on the book out of Barrett's *Voice* office, using computers, phones, faxes, and other facilities of the paper. Also contributing were other members of the *Voice* family: Tom Robbins, Jarrett Murphy, and Hervay Petion, as well as ex-Voicers Rob Morlino, Rashmi Vasisht, Laurence Pantin, and Eddie Borges.

The intern army falls into four categories. First, Annachiara Danieli carried the book alone on her shoulders at times, while Barrett was busy at the *Voice* and before Lenzer was hired, interviewing critical sources and collecting court and other records, particularly in 2003. Working without pay for many months, she laid the reporting foundation for this book. Christine Lagorio and Adam Hutton worked both as interns and as paid part-time assistants on aspects of the book over an extended period, and made substantial contributions. A large group of interns completed three-month intern cycles devoted almost entirely to book work: Richard Myers, Rema Rahman, Jane Lee, Kay Grigar, Annika Mengisen, Alex Gecan, Bryan Farrell, Stephen Stirling, Ian Kriegish, Nicole d'Andrea, Leslie Kaufmann, Tommy Hallissey, Andrew Burtless, Brian O'Connor, Erin Donar, Michael Mitchell, Ryan Vu, Anjuli Shukla, and Lyle Sclair. The final group split their time between *Voice* and book work: Jill Nawrocki, Jeff Herman, Solana Pyne, Clementine Wallace, Jess Wisloski, Lauren Johnston, Peter Madsen, Catherine Shu, Jennifer Suh, Andrea Toochin, Abby Aguirre, Marc Shultz, Daniel Magliocco, Ned Thimmayya, Ben Shestakofsky, Daniel Ten Kate, Caitlin Chandler, Emily Bond, Jessica Bennett, Lee Norsworthy, Xana O'Neill, and Ben James. The excellence and dedication of these extraordinary reporters was indispensable to the wide sweep of this book, even though some of the paths we took together did not wind up part of the final manuscript.

We have also been helped by many other journalists. None has influenced us more, particularly in the evacuation stories of the handicapped and those above the fire in the South Tower, than *102 Minutes*, the best seller by Jim Dwyer and Kevin Flynn. Dwyer and Flynn have also shared their time and thoughts with us again and again, not just as colleagues, but as collaborators who see their work and ours as linked chapters in the unfolding saga of one of America's darkest days. Dwyer, David Kocieniewski, Deidre Murphy, and Peg Tyre also wrote *Two Seconds Under the World*, the masterful source of much of our own description of the destruction of the 1993 World Trade Center bombing. Dwyer's and Flynn's articles in the *New York Times* were also helpful—on everything from the 9-1-1 breakdown to radios.

Journalists who have either written pieces that helped shape this book or whispered in our ear include: Al Baker, Eric Lipton, James Glanz, Ford Fessenden, Charles Bagli, Gail Collins, David Kocieniewski, Dan Barry, Ed Wyatt, and Willie Rashbaum of the *New York Times*; and Jesse Drucker, Rob Polner, Ti-Hua Chang, Russ Buettner, Bill Murphy, Richard Steir, Dan Janison, David Seifman, Joe Calderone, Graham Rayman, Sara Kugler, Alexander Stille, Leonard Levitt, Ellis Henican, Greg Sargent, Michael Daly, Juan Gonzalez, Paul

Thompson, and Elizabeth Kolbert. No 9/11 historian, journalist, or investigator can function without consulting the great work of Dennis Cauchon and Martha Moore of *USA Today*. Marty Rosenblatt, one of the best investigators in the city and an asset to countless reporters, found people for us no one else could find. Nick Von Hoffman was a professional shoulder to lean on.

Other helpful books were *Radio Silence FDNY* by Battalion Chief John Joyce and Bill Bowen, *American Ground* by William Langewiesche, *High Rise Fire and Life Safety* by former FDNY Fire Commissioner John T. O'Hagan, *1000 Years for Revenge* by Peter Lance, *The 1993 World Trade Center Bombing* by Charles J. Shields, *The Man Who Warned America* by Murray Weiss, *The Full Rudy* by Jack Newfield, *Report from Ground Zero* by Dennis Smith, *The Prince of the City* by Fred Siegel, *Out of the Blue* by Richard Bernstein, *Rudy! An Investigative Biography of Rudolph Giuliani* by Wayne Barrett, *Rudy Giuliani: Emperor of the City* by Andrew Kirtzman, *So Others Might Live* by Terry Golway, *American by Choice* by Captain Alfredo Fuentes, *City in the Sky* by James Glanz and Eric Lipton, *After* by Steven Brill, *My FBI* by Louis J. Freeh, *Never Forget* by Mitchell Fink and Lois Mathias, *Men of Steel* by Karl Koch III, *The Cell* by John Miller and Michael Stone, *Fallout* by Juan Gonzalez, *Leadership* by Rudolph W. Giuliani, *Lost Son* by Bernard Kerik, *The 1993 World Trade Center Bombing* by Peter Caram, and *Strong of Heart* by Thomas Von Essen.

While many sources cooperated with us, some require special thanks for supplying documents or giving substantial amounts of their time: Anthony Fusco, Lenora Gidlund, Jerry Hauer, Ray Kelly, Mary Jo White, Vincent Dunn, Glenn Corbett, Lou Anemone, Greg Semendinger, Paul Browne, Ilyse Fink, Bill Cunningham, Jordan Barowitz, Martin J. Steadman, Sally Regenhard, Monica Gabrielle, Lou Mangone, Victor Kovner, Richard McAllan, Andrew McCarthy, Gil Childers, David Kelley, Iris Weinshall, Fran Reiter, Bob Kerrey, John Lehman, Dietrich Snell, John Miller, John Elliff, John Hotis, Thomas Grady, Jim Margolin, Eric Lustig, Mike Arena, Kate Ahlers, Ruth Messinger, Danica Gallagher, John Odermatt, Cornelius Lynch, Michael Cohen, Zach Zahran, William Motherway, Warren Stogner, Andrew Carboy, Al

Fuentes, John Joyce, Richard Tell, Robert Schierenbeck, Rosaleen Tallon, Laura Weinberg, Tony Shorris, Richard Tomasetti, Alan Fitzer, Dot Carlson, Jeff Simmons, Kevin Culley, Ken Holden, Chris Marley, Joel Kupferman, Jonathan Bennett, Andrew Eristoff, Rudy Rinaldi, Ralph Balzano, Daniel Nigro, Steven Kuhr, Madeline Provenzano, Joe Lhota, Jimmy Boyle, Peter King, Tom Regan, Pete Caram, Ken Leibowitz, Andrew Burdess, Steve Savas, David Neustadt, Russ Pecunies, Richard Salem, Cate Jenkins, Congressman Jerry Nadler, Kathryn Freed, Norman Siegel, Ron Kuby, John Dyson, Wilbur Ross, Norman Steisel, Harvey Robbins, Phillip Wearne, Laura Harris, Doug Turetsky, Luke Healy, John Mohan, Nicole Gordon, Carl Meinhardt, Bob McGrath, Michael Stein, Corey Bearak, Robert Schwaninger, Steven Riegel, Wendy Niles, Erica Jurgel, Dr. Robin Herbert, Dan Greenberg, Pete Gorman, Barry Ostrager, David Rozenzweig, James O'Neill, Bob Caporale, Rick Mahoney, Kyle Krall, John Moran, Bruce Lippy, Dr. JoAnn Difede, H. Peter Haveles, and Darryl Fox.

And there were those who gave us great insight into the suffering and sacrifice that marked that terrible day and the rescue and recovery at Ground Zero that followed. So we offer thanks to Stephen King, Joe Libretti, Joe Firth, Dave O'Neal, Harold Schapelhouman, and Billy Trolan. Also, Dennis Gault, Tom Goodkind, and Catherine McVay Hughes.

The 9/11 Commission staff that worked on the chapter of the final report that dealt with the emergency response was also generous with their time and intelligence about these matters: John Farmer, Sam Caspersen, Cate Taylor, and George Delgrosso. Shyam Sunder and Michael Newman of the National Institute of Standards and Technology were also helpful. While the current administration of the Port Authority resisted the release of any public information, its executive directors during the pre 9/11 years, Stan Brezenoff, George Marlin, and Robert Boyle, were all willing to fully explore this history, as was Mel Schweitzer, then a member of the authority's board. Some city officials, the Fire Department's Frank Gribbon and Alex Fisher, city attorney Gary Shaffer, and Agostino Cangemi at the Department of Information Technology and Telecommunications provided some documents.

NOTES

This is a mix of traditional footnotes and overview source notes. The source notes cut across chapters. For example, source note 1 describes the various sources for quotes from Rudy Giuliani, Bernard Kerik, Richard Sheirer, and Tom Von Essen. The first time each of these four is quoted, a summary of source note 1 is cited again. Most of the rest of the quotes from the four do not refer to a source note. There are several other source notes like 1 that cover multiple sources or frequently cited material. We do not footnote them every time we use them. If a quote comes from a source interviewed by the authors, it usually does not get a source note. A source note is used with an interviewed subject only in instances when the reader might be confused about whether or not the quote comes from an interview. If the material is adequately sourced in the text, it also does not get a source note. The source notes do not attempt to cover every new fact in the book, only those that appear to require deeper sourcing or explanation.

CHAPTER 1

1. Rudy Giuliani, Tom Von Essen, Richard Sheirer, and Bernard Kerik, all of whom were employed at Giuliani Partners for most of the last four years, declined to be interviewed. Their frequent quotes are drawn from media sources, as well as Giuliani's book *Leadership* (Miramax, 2002), Von Essen's book *Strong of Heart* (HarperCollins, 2002), and Bernard Kerik's book *Lost Son* (HarperCollins, 2001). Some quotes are drawn from speeches that we obtained or covered—for example, Von Essen's appearance at a Las Vegas convention for Motorola or

Giuliani's question-and-answer session with Mexico City businessmen. Von Essen, Sheirer, and Kerik also testified at a public hearing of the 9/11 Commission on May 18, 2004, and Giuliani on May 19. In addition to their public testimony, Giuliani testified privately before commission staff on April 20, 2004, Von Essen on April 7, Kerik on April 6, and Sheirer on April 7. The memorandums for the record (MFRs) of the confidential commission interviews are not verbatim transcripts but, rather, notes marked by commission staff as "100 percent accurate." Usually, where quotation comes from the private testimony, it is described in the text as a statement made to commission staff or investigators. Internal commission documents prepared before the public testimony of each of these men are also sometimes quoted.

2. Zack Zahran was interviewed by the authors.

3. Eric Levine was interviewed on the BBC *Eyewitness* series.

4. The National Institute of Standards and Technology, a subdivision of the U.S. Department of Commerce, is still conducting a study examining the collapse of 1, 2, and 7 World Trade Center. Its final report on 1 and 2 WTC, issued in 2005, was over 10,000 pages long. Its preliminary reports were also thousands of pages. Much the evacuation data in the book is drawn from its studies, and it is cited on issues ranging from the radios to command and control. We have attended its periodic New York briefings, interviewed several members of the investigating team, and visited its headquarters in Maryland, reporting on a two-day public session there. Its press representative, Michael Newman, has responded in writing to most of our e-mailed questions. We have also interviewed the director of the study, Shyam Sunder.

5. Much of the information about President Bush's movements on 9/11 is taken from Scott Pelley's interview with the president, *60 Minutes II*, September 10, 2003. Some of the details, including precise times, are derived from many other media sources.

6. Christopher Anderson's *George and Laura: Portrait of an American Marriage* (HarperCollins, 2002).

7. Giuliani, Von Essen, Sheirer, and Kerik documents and testimony.

8. Jonathan Alter, *Newsweek*, September 27, 2001.

9. "Person of the Year," *Time*, December 31, 2001.

10. Tom Roeser, *Chicago Sun-Times*, September 10, 2005.

CHAPTER 2

1. Sam Caspersen granted a four-hour interview and requested that he be allowed to review the typed-up notes of the interview. When he got the notes, he responded with a detailed, 18-page, point-by-point assessment, and in some cases, revision of his comments. He is quoted frequently here, either from the interview notes or from his written response. We have tried to include any caveats he added in the text, but it was not always possible. For example, his letter confirmed his belief that OEM did not have a direct impact on the rescue/evacuation operation, but added that one OEM official "was contacting hospital officials

in order to gear them up for what was thought to be the imminent arrival of thousands of injured civilians and first responders." Caspersen says that the OEM staff at the command center also contacted 20 predesignated representatives of different agencies "and asked them to respond asap to OEM headquarters." They were en route to the center, Caspersen says, when it was evacuated at 9:30.

2. Paul Browne is referring to the brilliant documentary film made by Jules and Gedeon Naudet, shot mostly in the North Tower lobby. It is also the source of the Sheirer statement to Von Essen about another plane.

3. Giuliani, Von Essen, Sheirer, and Kerik documents and testimony.

4. Naudet documentary.

5. Giuliani, Von Essen, Sheirer, and Kerik documents and testimony.

6. Memorandums for the record (MFRs), written by 9/11 Commission staff after private interviews, are also cited in the text for Edward Plaugher, the Arlington County fire chief, taken on October 16, 2003; Alan Reiss, the World Trade Center director for the Port Authority, November 3, 2003; and Jerome Hauer, former director of the Office of Emergency Management, March 30, 2004. All three also testified at the commission's public hearings in May 2004.

7. A high-ranking executive in the mayor's office recalled former police commissioner Safir using this term at a critical meeting about the command center, but the executive declined to be identified by name. Direct calls to Safir's security consulting company, and a written request to Safir publicist Howard Rubenstein, did not result in an interview.

8. *Never Forget*, by Mitchell Fink and Lois Mathias (HarperCollins, 2002), contains oral history interviews of Joe Pfeifer, Jay Jonas, Tom Von Essen, Bernard Kerik, Joe Esposito, Billy Butler, and Hector Santiago that are cited at various points in this text.

9. John Peruggia's oral history was taken by the Fire Department on October 25, 2001.

10. Giuliani, Von Essen, Sheirer, and Kerik documents and testimony.

11. Jerry Hauer, the first director of the Office of Emergency Management, did five extended interviews and also supplied the agendas, correspondence, and other documents related to his meetings with Mayor Giuliani from 1996 through 1999. Lou Anemone, the chief of the department at the NYPD, granted an extensive interview and supplied detailed records from the NYPD exchanges with OEM regarding command and control issues. Both turned over copies of proposed protocols and matrixes.

12. The World Trade Center Task Force report was issued in March 1995 by Fire Commissioner Howard Safir and Department of Buildings Commissioner Joel Miele. The study was initially authorized in March 1993 by fire commissioner at the time, Carlos Rivera, and buildings commissioner at the time, Rudolph Rinaldi.

13. In addition to preparing memos for departmental use, Chief Anthony Fusco wrote and organized a lengthy overview of the 1993 response that was initially published in *WNYF*, the official publication of the FDNY. After that ex-

tended review was published in September 1993, Fusco and colleagues also published another compendium of articles in *Fire Engineering* magazine in December 1993.

14. Philip Zelikow appeared on National Public Radio on September 2, 2004.

15. The Mongello and Larsen quotes are taken from *102 Minutes* by Jim Dwyer and Kevin Flynn (Times Books, 2005), and their work in the *New York Times*, as well as the book, has informed this section on Stairway A.

16. Sheirer was elevated to assistant commissioner at the Fire Department while David Dinkins was mayor, so he could not openly participate in the Giuliani campaign in 1993. The commissioner who promoted him, Carlos Rivera, wound up endorsing Giuliani and resigning during the campaign.

17. The Chris Young story is derived from Dwyer and Flynn's book, *102 Minutes*. In addition to the *USA Today* stories, the Fusco after-action reports, FDNY elevator expert John Hodgens, *Elevator World* editor Bob Caporale, the National Institute of Standards and Technology report, and chief mechanic James O'Neill contributed to this account.

18. John Lehman said this in an interview with the authors.

19. Sam Caspersen said that the Safety division "is a backwater where black sheep are put within the department" and that "officers come out of the ranks with no special background." He also said there's no written standard operating procedure for the division and that "new officers ride around with an old chief and pick up orally how to do the job."

20. Peter Lance, *1000 Years for Revenge* (HarperCollins, 2003), which recounts the extraordinary story of Bucca, who made it to the 78th floor of the South Tower and died in the collapse.

21. The National Institute of Standards and Technology came up with these numbers, though others question them.

22. Jim Dwyer and Kevin Flynn, in *102 Minutes*, described the use of these evacuation chairs and only cited instances when Port Authority employees were brought down in them. Dwyer and Flynn also initially reported the Beyea/Zelmanowitz story.

23. Fink and Mathias, Never Forget.

24. Zelikow interview, NPR, September 2, 2004.

25. Dwyer and Flynn reported Walsh's warning.

26. Port Authority police chief Morris was quoted in *Newsday*, January 23, 2002.

27. Alfredo Fuentes, *American by Choice* (Fire Dreams Publishing, 2004).

CHAPTER 3

1. In addition to court records and U.S. Senate and House of Representatives hearing transcripts, the description of the 1993 bombing is derived from "Two Seconds Under the World," by Jim Dwyer et al. (Crown Publishers, 1994),

and *The 1993 World Trade Center Bombing*, by Charles Shields (Chelsea House, 2001).

2. A campaign aide who requested that he not be identified described the Giuliani trip toward the trade center in February 1993. Another campaign aide recounted how they ridiculed Dinkins's lack of leadership in the aftermath of the 1993 bombing.

3. The authors interviewed Pete Caram.

4. The authors have reviewed a tape of this session; this is also Bratton's memory as relayed by John Miller, who was his spokesman for most of the time that Bratton was NYPD commissioner. At the time of a two-hour interview with Miller in 2005, he was Bratton's spokesman at the Los Angeles Police Department. Miller is currently the chief spokesman for the FBI. His closeness to Bratton, even after Miller left the NYPD, continued for at least a decade, and Miller said he could speak for Bratton on this history. Bratton's own autobiography, *Turnaround* (Random House, 1998), contains no reference to terrorism, though it is largely an account of his NYPD years.

5. This total is drawn from internal law firm documents. A *Newsday* story in 1993 investigated some of these questionable clients and revealed the 1991 memo.

6. Fitzgerald is the only one named who did not talk to us. But David Kelley and Mary Jo White, both of whom served as U.S. attorney in the Southern District and were very close to Fitzgerald, said they did not believe Fitzgerald ever briefed or was asked to brief Giuliani or anyone else at City Hall.

7. An FBI spokesman, a commission source, the NYPD's Browne, and Chief of Department Lou Anemone also indicated they believe the number of detectives assigned to the terrorism task force remained largely unchanged, at a level of 16 or 17, throughout the Giuliani years.

8. The Schwartz documents, as well as many other documents cited throughout this book, were obtained from the Giuliani archive maintained by the New York City Department of Records and Information Services.

CHAPTER 4

1. Jerry Hauer, Lou Anemone, and John Miller confirmed that Freeh personally briefed Giuliani on specific terrorist events or threats.

2. The three executive directors of the authority for almost all of the Giuliani years, Stan Brezenoff, George Marlin, and Robert Boyle, were interviewed and agreed about this. Top deputies at the authority, including Tony Shorris and others who asked not to be identified, said the same thing.

3. George Marlin, executive director of the Port Authority, provided access to many key authority documents covering the 1994–1997 period.

4. Laura Weinberg, whose husband, Richard Aronow, worked in the Port Authority legal department before dying in an elevator, where he was trapped on his way to work.

5. Savas and Steisel recalled these discussions.

6. A comprehensive search of the Giuliani records maintained at the municipal archives uncovered no other reference to the impact of the 1993 bombing on the trade center or Port Authority.

7. Boyle and McLaughlin recounted these contacts.

8. Tom Robbins revealed this chat room conversation in a series of *Village Voice* stories that eventually led to Harding's indictment and conviction.

9. Rob Polner, a *Newsday* reporter, saw Dyson leaving with the Rolodex.

10. The authors submitted a freedom-of-information request in 2003 to the Department of Buildings requesting all records related to 1, 2, and 7 WTC. Over the course of three visits to the department, we reviewed thousands of department documents. We did the same with the Fire Department.

11. Ibid.

12. The National Institute of Standards and Technology documented the lack of refireproofing that occurred in the towers, as well as the yearly pace of the project. It also established the effect of primer paint on the adhesive strength of the fireproofing.

13. NIST reported that the fourth stairway was not raised at this meeting.

CHAPTER 5

1. 9/11 Commission staff memorandums for the record (MFRs).

2. Ibid.

3. Hauer and Anemone interviews and documents.

CHAPTER 6

1. In response to a freedom-of-information request, the Department of Citywide Administrative Services, the Office of Management and Budget, the Department of City Planning, and the Corporation Counsel of the City of New York turned over hundreds of documents related to the rental and construction of space at 7 World Trade Center for the Emergency Operations Center. Con Ed and its attorneys also supplied some records. The Office of Emergency Management, which trained all city agencies in how to back up records particularly with regard to Y2K, claimed it lost all its records on 9/11. Though city officials were quoted in the *New York Times* in 2002 as saying that they'd received a Justice Department grant to do an after-action report on OEM's response that day, adding that the report would be completed soon, a city attorney said that he believed no such report exists, and OEM stopped returning messages.

2. Glenn Pymento and Richard Ramos were deposed as part of the lawsuit brought against the city by Con Edison and its insurers, as was Hauer.

3. Silverstein's daughter married her partner in a televised ceremony and wrote a book with her partner.

4. Pymento and Ramos deposition.

5. A high-ranking executive in the mayor's office quoting Safir, who declined to be interviewed.

CHAPTER 7

1. City Comptroller William Thompson, in response to several freedom-of-information requests, and the Fire Department turned over most of the documents cited in this chapter. Some, particularly those involving the Department of Information Technology and Telecommunications, were obtained from the municipal archives. Warren Stogner of General Electric/Ericsson also supplied some records.

2. Jack Newfield and Wayne Barrett, *City for Sale: Ed Koch and the Betrayal of New York* (HarperCollins, 1989).

3. Steven Gregory, Don Stanton, and Robert Scott, the three FDNY officials who are still at the department and were involved with the radio decisions, declined to be interviewed. Another decision maker, Tom Fitzpatrick, who is now with Giuliani Partners, also declined to be interviewed. Stanton's father, who was involved in radio purchases for Emergency Medical Services, could not be located.

4. Ibid.

5. *The Chief*, a newspaper for municipal workers, reported on this meeting in June 2001.

6. The Department of Information Technology and Telecommunications, through its attorneys, maintains that no records of Deborah Spandorf's seven years at the department exist, explaining that she communicated mostly by e-mail.

7. At the end of an extended telephone interview, Marian Barell suggested we call again after we completed our reporting. Asked about why she'd moved out to Illinois, so close to Motorola headquarters, after retiring from Motorola, she simply insisted she had nothing to do with Motorola. She did not mention that she had remarried. Her new husband, Robert Barnett, a top Motorola executive, answered the phone when we called again and said that Motorola's national press office would handle any further questions, including any concerning his marriage. When Steve Gorecki, the press officer for Motorola, was asked if it was usual for the company to handle personal press questions about an ex-employee like Barell, who'd left years earlier, he said no. Detailed questions were e-mailed to Gorecki about a wide range of Motorola matters, as well as the delicate timing questions of the Barnett/Barell relationship, and the company declined to answer them. Barnett's former wife, Kathleen, said she knew nothing about when her former husband had met Marian Barell, adding that she did not have any information about another woman in her ex-husband's life when she sought the divorce. Nor has she learned of any since, she said.

8. Harte declined to be interviewed. Several calls to Ahmed and Weinberg at their Motorola offices, and at Weinberg's home, went unreturned. Gino Menchini, the most recent DOITT commissioner, agreed to an interview and then canceled.

9. Though detailed questions about this possible conflict were sent to Arnold & Porter, they got back to us and specifically declined to answer any of them.

CHAPTER 8

1. Gail Collins, "New York Notes 8 Million Survivors in Need of Affection," *New York Times*, September 17, 2001.

2. Juan Gonzalez, "Fallout," *The New Press*, 2002.

3. William Langewiesche's *American Ground: Unbuilding the World Trade Center* (North Point, 2002) was a critical guide to life at Ground Zero.

4. Stephanie Armour, "Health Problems Plague Ground Zero Workers," *USA Today*, March 3, 2003.

5. On September 29, the mayor suggested that the rescue phase of the operation was over, but no change was made in the actual management of the site. Giuliani said that the reality was that the city no longer "expected to find anybody alive," but added that the city would "conduct the operation in the same way" because "we are doing the same thing we would do to recover someone alive as to recover human remains." The Fire Department remained the incident commander at the site, running the operation. It wasn't until October 31 that the Department of Design and Construction (DDC) was nominally named the coincident commander with the FDNY, and testimony in the pending Ground Zero lawsuits indicates that even that appointment did not change site management. The search for body parts, overseen by the FDNY with separate NYPD and Port Authority units conducting their own operations, was in effect a continuation of the rescue phase. Giuliani told Fox News, "It really doesn't matter if you describe it as a rescue effort." In Von Essen's book, he indicates that the conversion to a recovery/demolition/cleanup/reconstruction phase never happened, and, in an interview, DDC Commissioner Ken Holden agreed. Holden also testified in a pending Ground Zero lawsuit that "no one ever told me I was an incident commander and I don't know what an incident commander is." Holden's deputy Mike Burton added in his own court deposition in 2005 that a deputy mayor did tell him that DDC was to serve as co-incident commander. He said that all it meant was that he and a fire chief held joint daily meetings rather than sending representatives to the separate meetings they previously hosted.

6. Maggie Haberman, "Hizzoner Hits Hardhat Holiday Gripe," *New York Post*, November 24, 2001.

7. Dr. Philip Landrigan wrote an article in the *American Journal of Industrial Medicine* called "Lessons Learned: Worker Health and Safety Since September 11, 2001," published in 2002.

8. Dr. Michael Weiden read a statement in October 2003 that he jointly prepared with Dr. Kerry Kelly and Dr. David Prezant for the Congressional Subcommittee on National Security, Emerging Threats and International Relations. Kelly is the FDNY's chief medical officer and Prezant and Weiden are top assistants in the department's Bureau of Health Services.

9. Bruce Lippy, "Respiratory Protection at the World Trade Center: Lessons From the Other Disaster," The National Clearinghouse for Worker Safety and Health Training. www.wetp.org/wetp/wtc/Respiratoruse.pdf

10. Juan Gonzalez, "A Toxic Nightmare at Disaster Site," New York *Daily News*, October 26, 2001.

11. Francesca Lyman, "Messages in the Dust: What Are the Lessons of the Environmental Health Response to the Terrorist Attacks of September 11?," National Environmental Health Association publication, September 2003.

12. In a January 2004 interview with EPA regional communications director Bonnie Bellow, she said that Whitman "felt so strongly about the need for the use of respirators" that she made these midnight calls "while watching CNN covering Ground Zero workers without respirators on."

13. Michelle Goldstein memo to Deputy Mayor Robert Harding, re: "Legislative Alternatives to Limit the City's Liability relating to 9/11/01," undated.

14. Bruce Lippy, "Cleaning Up After 9/11: Respirators, Power and Politics," *Occupational Hazards*, May 29, 2002.

15. William Langewiesche, "American Ground."

16. As quoted in *Never the Same*, a documentary by Jonathon Levin.

17. Ginger Adams Otis, "Assembly Seeks to Help Those Injured on 9/11," *The Chief*, September 2, 2005.

18. Final Report of the Special Master for the September 11 Victim Compensation Fund, undated.

19. Gary Shaffer, New York City assistant corporation counsel, supplied these numbers, as the city prepared papers to file in these cases.

20. Anthony DePalma, "Many Who Served on 9/11 Press Fight for Compensation," *New York Times*, May 13, 2004.

21. Ridgely Ochs, "Ailments, Struggles of 9/11 EMT Who Died Not Unique," *Newsday*, September 4, 2005.

CHAPTER 9

1. *Late Show with David Letterman*, CBS transcript, September 17, 2001.

2. Sam Smith, "Furor over WTC Lies," New York Post, July 18, 2004. The Post article relied on the findings of the New York Law and Justice Project, which posted its full freedom of information response from the New York State Department of Environmental Conservation on its website between January 22, 2002 and June 4, 2004. This posting revealed the unreleased results of the New York City Department of Environmental Protections tests. These findings are now frequently cited in the litigation against the city and the EPA, the federal agency that coordinated the DEP and all other air and dust sampling.

3. "EPA's Response to the World Trade Center Collapse: Challenges, Successes, and Areas for Improvement," Office of Inspector General, Report No. 2003-P-00012, August 21, 2003.

4. Andrew Schneider, "NY Officials Underestimate Danger," St. Louis Post-Dispatch, January 13, 2002.

5. The EPA Inspector General's report is the primary source for much of the asbestos and toxin data here.

6. Juan Gonzalez, *Fallout* (New Press, 2002).

7. New York Committee for Occupational Safety and Health, submitted to the City Council, March 2002.

8. Lyman, "Messages in the Dust."

9. David Newman, an industrial hygienist with NYCOSH, quoted in "Cleaning Up After 9/11: Respirators, Power and Politics," *Occupational Hazards*, May 29, 2002.

10. The EPA IG report described the city's instructions to the public.

11. Lyman, "Messages in the Dust."

12. The EPA IG report contained the city's response, submitted by Kenneth Becker, an assistant corporation counsel. The report also contained the FEMA statements about the city's position on a cleanup program.

13. Ibid.

14. Gonzalez, *Fallout*.

15. Muszynsky was interviewed by the authors.

16. Lyman, "Messages in the Dust."

17. The same day that Judge Batts held Whitman personally liable, another judge in the same courthouse, Alvin Hellerstein, declined to find her liable in a case with some similarities to the Batts case. Both decisions are currently on appeal.

18. Sally Ann Lederman, et al., "The Effects of the World Trade Center Event on Birth Outcomes among Term Deliveries at Three Lower Manhattan Hospitals," *Environmental Health Perspectives*, December 2004. Anthony M. Szema, et al., "Clinical Deterioration in Pediatric Asthmatic Patients After September 11, 2001," *Journal of Allergy and Clinical Immunology*, March 2004.

19. Shao Lin, et al., "Upper Respiratory Symptoms and Other Health Effects Among Residents Living Near the WTC Site After September 11, 2001," *American Journal of Epidemiology*, September 15, 2005.

20. Lyman, "Messages in the Dust."

21. Kristen Lombardi, "Dusted: Long after 9/11, Some People Say the Dust is Still Making Them Sick," *Village Voice*, September 6, 2005.

CHAPTER 10

1. Jack Newfield, *The Full Rudy: The Man, the Mayor, the Myth* (Nation Books, 2003).

2. William Cunningham, communications director for Mayor Bloomberg, said the mayor and ex-mayor talked just a couple of times during the first few months of 2002. Bloomberg also didn't feel compelled to continue Giuliani initiatives. The *New York Times* noted that he eliminated the Vacancy Decontrol Board, which Giuliani had used to control city hiring; axed a deal to sell a building near the United Nations; and was reconsidering a lease of city land in Brooklyn to one of Giuliani's political supporters. A spokesman called the decisions "a reflection of changing needs and priorities in difficult times."

3. Giuliani's loyalty to old associates extended beyond City Hall. One of his oldest friends, Father Alan Placa, was employed by Giuliani Partners after he was suspended as a priest during the child molestation scandal.

4. Michael Hess told the *New York Times* in December 2004 that Bear Stearns was not working out, and the *Times* said the relationship "might soon be abandoned" (Eric Lipton, December 4, 2004).

5. Edward Iwata, "Giuliani's Firm's Deal to Advise Company Raises Questions," *USA Today*, March 17, 2005.

6. Eric Lipton of the *New York Times* mentions this incident in a February 22, 2004, story, which was one of the few serious analyses of Giuliani Partners. Many of the points Lipton made are reflected here.

7. Laurence Pantin, a Mexico City–based reporter, covered the videoconference for us and taped the exchange.

8. Daniela Gerson reported the negative response to Giuliani Partners in the *New York Sun*, April 11, 2005.

9. Ben Smith, "Really Rich Rudy," *New York Observer*, April 4, 2005.

10. While Dempsey credited Nextel with "making their best effort" to respond to NYPD complaints, that clearly wasn't the case nationally. A top Nextel executive said in 2003 that over the years, the company had resolved only 30 of some 700 interference complaints across the country. Dempsey also said that the city's interagency frequency, the 800-megahertz system set up by the Office of Emergency Management, started to experience interference in the late '90s. "The FCC suggested they buy new equipment," Dempsey recalled. The NYPD was experiencing this Nextel interference even though its radios were operating on 470 megahertz. Dempsey said the interference was "strictly power" and that Nextel sites were "saturating our receivers with pure signal."

11. On September 16, 2001, Roz Allen, an attorney with Arnold & Porter representing the city, sent an e-mail to the FCC saying that "Nextel agreed to coordinate cells on wheels (COWs) and other temporary base stations with the city and the utilities," adding that "without this coordination, it is highly probable that Nextel would disrupt these other critical communications." This indicated that the 7,000 phones Nextel delivered to Ground Zero immediately after 9/11 were a cause of concern to the city because of potential interference with other public safety communications there. The same concerns about phones like Nextel's prompted Agostino Cangemi, the DOITT general counsel, to testify at a Senate Subcommittee on Communications hearing in March 2002 that the city was monitoring the 800-megahertz frequencies closely in the immediate aftermath of 9/11. This monitoring, said Cangemi, "was especially important because of the many temporary wireless facilities, including COWs, being deployed."

12. "I cannot imagine a heavier time," a Nextel official said of its September 11 call volume in news accounts the next day.

13. The city began filing these interference complaints only in 2002, partly in response to the Nextel spectrum proposal. Clearly, as Dempsey says, interfer-

ence was occurring prior to 2002, but the city wasn't filing complaints. In a 2002 submission to the FCC, the city said: "New York City's public safety frequencies are plagued with interleaving and interference problems caused by commercial carriers. The record to date strongly suggests that while a channel reshuffling may well mitigate interference, reshuffling alone is unlikely to eliminate interference." Since this submission was a comment filed with the Nextel spectrum application, it was a clear reference to the unique interleaving effect Nextel had on public safety communications. This interleaving occurs because Nextel frequencies were literally mixed in with police and fire frequencies.

14. The half billion dollars in public safety rebanding costs was the initial offer, and is expected to grow to more than a billion. In addition, Nextel is doing rebanding that affects private users. The agreement with the FCC provides that whatever Nextel doesn't spend on rebanding or get FCC credit for, it must pay to the government, up to the estimated value of the new spectrum. Opponents of the plan have charged that the true value of the new spectrum is greater than the estimate. Since this deal was done without the usual public auction, the market did not determine the spectrum's value.

15. *Dispatch Monthly*, 2002 coverage of the APCO national convention.

16. Tony Carbonetti, a Giuliani political aide at City Hall with no known background in public safety communications, actually became the chair of a subcommittee of the FCC's Media Security and Reliability Council. Fitzpatrick was a member of another subcommittee.

17. Eric Lipton, in the *New York Times* on February 22, 2004, reported the $15 million in stock options.

18. Within months of the FCC approval of the Nextel deal, Sprint acquired it. Sprint had filed papers with the FCC opposing Nextel's plan, pointing out that the "primary causer of interference" would be rewarded with "an unwarranted spectrum grab."

19. Melinda Ligos, "Private Sector: Moving in on New York Lawyers," *New York Times*, February 15, 2004.

20. "At Home with Judith Giuliani," *Avenue*, November 2003.

21. Thomas Maier, "Adviser to Ridge, Adviser to Business Partners," *Newsday*, December 11, 2004.

CHAPTER 11

1. Craig Bildstien, "The Cherie Effect," *The Advertiser* (Adelaide, Australia), February 10, 2005.

2. Suzanne Wilson, "Giuliani Dinner Fails to Meet Charity Goal," *Calgary Herald*, October 25, 2002.

3. "The Note," *ABC News*, March 31, 2005.

4. Associated Press, April 2, 2005.

5. "Rudy and the Right," *New York Post*, February 10, 2006.

6. Wayne Barrett, "Romancing the Right," *Village Voice*, February 15, 2000.

7. "What Price Success?," *Tampa Tribune*, February 24, 2002.

8. Joyce Purnick, "Busy Being an Icon," *New York Times*, September 5, 2002.
9. Raleigh *News & Observer*, November 4, 2003.
10. *Houston Chronicle*, March 30, 2005.

CHAPTER 12

1. Kevin Culley, the fire captain assigned to OEM, said they heard over the radio that the eagle was heading in and they got a laugh out of the code name for Giuliani.

2. Esposito and Dunne have described their own movements in interviews; Esposito's was published in *Never Forget* by Mitchell Fink and Lois Mathias.

3. Bernard Kerik's private testimony before the 9/11 Commission staff, April 6, 2004.

4. 9/11 Commission report.

5. The National Institute of Standards and Technology found that there were only 17,400 people in the towers, with approximately 15,000 surviving. It also determined that almost all the survivors began their evacuations within three minutes of the crash in the North Tower and six minutes in the South Tower.

INDEX